Neurophilosophy at Work

In this collection of essays, Paul Churchland explores the unfolding impact of the several empirical sciences of the mind, especially cognitive neurobiology and computational neuroscience, on a variety of traditional issues central to the discipline of philosophy. Representing Churchland's most recent investigations, they continue his research program, launched more than thirty years ago, which has evolved into the field of neurophilosophy. Topics such as the nature of consciousness, the nature of cognition and intelligence, the nature of moral knowledge and moral reasoning, neurosemantics or "world representation" in the brain, the nature of our subjective sensory qualia and their relation to objective science, and the future of philosophy itself are here addressed in a lively, graphical, and accessible manner. Throughout the volume, Churchland's view that science is as important as philosophy is emphasized. Several of the colored figures in the volume will allow readers to perform some novel phenomenological experiments on their own visual system.

Paul Churchland holds the Valtz Chair of Philosophy at the University of California, San Diego. One of the most distinguished philosophers at work today, he has received fellowships from the Andrew Mellon Foundation, the Woodrow Wilson Center, the Canada Council, and the Institute for Advanced Study in Princeton. A former president of the American Philosophical Association (Pacific Division), he is the editor and author of many articles and books, most recently *The Engine of Reason, the Seat of the Soul: A Philosophical Journey into the Brain* and *On the Contrary: Critical Essays, 1987–1997* (with Patricia Churchland).

Neurophilosophy at Work

PAUL CHURCHLAND

University of California, San Diego

CAMBRIDGE
UNIVERSITY PRESS

CAMBRIDGE UNIVERSITY PRESS
Cambridge, New York, Melbourne, Madrid, Cape Town, Singapore, São Paulo

Cambridge University Press
32 Avenue of the Americas, New York, NY 10013-2473, USA

www.cambridge.org
Information on this title: www.cambridge.org/9780521864725

© Paul Churchland 2007

First published 2007

Printed in the United States of America

A catalog record for this publication is available from the British Library.

Library of Congress Cataloging in Publication Data
Churchland, Paul M., 1942–
Neurophilosophy / Paul Churchland.
p. cm.
Includes bibliographical references and index.
ISBN 0-521-86472-0 (hardback) – ISBN 0-521-69200-8 (pbk.)
1. Neurosciences – Philosophy. 2. Cognition – Philosophy.
3. Religion and science. I. Title.
QP356.C485 2007
612.801–dc22 2006014487

ISBN 978-0-521-86472-5 hardback
ISBN 978-0-521-69200-7 paperback

Contents

Preface

Any research program is rightly evaluated on its unfolding ability to address, to illuminate, and to solve a broad range of problems antecedently recognized by the professional community. The research program at issue in this volume is cognitive neurobiology, a broad-front scientific research program with potential relevance to a considerable variety of intellectual disciplines, including neuroanatomy, neurophysiology, neurochemistry, neuropathology, developmental neurobiology, psychiatry, psychology, artificial intelligence, and . . . philosophy. It is the antecedently recognized problems of this latter discipline in particular that constitute the explanatory challenges addressed in the present volume. My aim in what follows is to direct the light of computational neuroscience and cognitive neurobiology – or such light as they currently provide – onto a range of familiar philosophical problems, problems independently at the focus of much fevered philosophical attention.

Some of those focal problems go back at least to Plato, as illustrated in Chapter 8, where we confront the issue of how the mind grasps the timeless structure underlying the ephemeral phenomena of the perceivable world. And some go back at least to Aristotle, as illustrated in Chapters 3 and 4, where we confront the issue of how the mind embodies and deploys the moral wisdom that slowly develops during the social maturation of normal humans. Other problems have moved into the spotlight of professional attention only recently, as in Chapter 1, where we address the ground or nature of consciousness. Or as in Chapter 7, where we address the prospects of artificial intelligence. Or as in Chapter 9, where we confront the allegedly intractable problems posed by subjective sensory qualia. But all of these problems look interestingly different when viewed from the perspective of recent developments in the empirical/theoretical research program of cognitive neurobiology. The low-dimensional 'box canyons', in which conventional philosophical approaches have become trapped, turn out to be embedded within higher dimensions of doctrinal possibility, dimensions in which specific directions of development appear both possible and promising. Once we have freed ourselves from the idea that cognition is basically a matter of manipulating sentence-like states (the various 'propositional attitudes' such as perceives-that-P, believes-that-P, suspects-that-P, and so on), according to rules of deductive and inductive inference, and once we have grasped the alternative modes of world representation, information coding, and information processing displayed in all terrestrial brains, each of the problems listed earlier appears newly tractable and potentially solvable.

The distributed illumination here promised is additionally intriguing because it comes from a single source – the vector-coding and vector/matrix-processing account of the brain's cognitive activity – an empirically based account of how the brain represents the world, and of how it manipulates those representations. Such a 'consilience of inductions', as William Whewell would describe it, lends further credence to the integrity of the several solutions proposed. The solutions proposed are not 'independent' solutions: they will stand, or fall, together.

As the reader will discover, all but one of the essays here collected were written in response, either explicit or implicit, to the published researches of many of my distinguished academic colleagues,[1] and each embodies my attempts to exploit, expand, and extend the most noteworthy contributions of those colleagues, and (less often, but still occasionally) to resist, reconstruct, or subvert them. Though cognitive neurobiology hovers always in the near background, the overall result is less a concerted argument for a specific thesis, as in a standard monograph, but more a many-sided conversation in a parlor full of creative and resourceful interlocutors. To be sure, my voice will dominate the pages to follow, for these are my essays. But the voices of my colleagues will come through loud and clear even so, partly because of their intrinsic virtues, and partly because the point of these essays is to try to address and answer those voices, not to

---

[1] The exception is Chapter 5, the essay on American educational policy, specifically, on the antiscience initiatives recently imposed, and since rescinded, in Kansas. I had thought these issues to be safely behind us, but after the 2004 elections, fundamentalist initiatives are once again springing up all over rural America, including, once again, poor Kansas. The lessons of this particular essay are thus newly germane.

muffle them. Without those voices, there would have been no challenges to answer, and no essays to collect.

The result is also a journey through a considerable diversity of philosophical subdisciplines, for the voices here addressed are all in hot pursuit of diverse philosophical enthusiasms. In what follows, we shall explore contemporary issues in the nature of consciousness itself, the fortunes of nonreductive materialism (specifically, functionalism) in the philosophy of mind, the neuronal basis of our moral knowledge, the future of our moral consciousness, the roles of science and religion in our public schools, the proper cognitive kinematics for the epistemology of the twenty-first century, the basic nature of intelligence, the proper semantic theory for the representational states of terrestrial brains generally, the fortunes of scientific realism, recent arguments against the identity theory of the mind–brain relation, the fundamental differences between digital computers and biological brains, the neuronal basis of our subjective color qualia, the existence of novel – indeed, 'impossible' – color qualia, and the resurrection of objective colors from mere 'secondary' properties to real and important features of physical surfaces. What unites these scattered concerns is, once more, that they are all addressed from the standpoint of the emerging discipline of cognitive neurobiology. The exercise, as a whole, is thus a test of that discipline's systematic relevance to a broad spectrum of traditional philosophical issues. Whether, and how well, it passes this test is a matter for the reader to judge. My hopes, as always, are high, but the issue is now in your hands.

Provenances

"Catching Consciousness in a Recurrent Net," first appeared in A. Brook and D. Ross, eds., *Daniel Dennett: Contemporary Philosophy in Focus,* pp. 64–81 (Cambridge: Cambridge University Press, 2002).

"Functionalism at Forty: A Critical Retrospective," first appeared in *Journal of Philosophy* 102, no. 1 (2005): 33–50.

"Toward a Cognitive Neurobiology of the Moral Virtues," first appeared in *Topoi* 17 (1998): 1–14, a special issue on moral reasoning.

"Rules, Know-How, and the Future of Moral Cognition," first appeared in *Moral Epistemology Naturalized,* R. Campbell and B. Hunter, eds., *Canadian Journal of Philosophy,* suppl. vol. 26 (2000): 291–306.

"Science, Religion, and American Educational Policy," first appeared in *Public Affairs Quarterly* 14, no. 4 (2001): 279–91.

"What Happens to Reliabilism When It Is Liberated from the Propositional Attitudes?" first appeared in *Philosophical Topics,* 29, no. 1 and 2 (2001): 91–112, a special issue on the philosophy of Alvin Goldman.

"On the Nature of Intelligence: Turing, Church, von Neumann, and the Brain," first appeared in S. Epstein, ed., *A Turing-Test Sourcebook,* ch. 5 (The MIT Press 2006).

"Neurosemantics: On the Mapping of Minds and the Portrayal of Worlds," first appeared in K. E. White, ed., *The Emergence of Mind,* pp. 117–47 (Milan: Fondazione Carlo Elba, 2001).

"Chimerical Colors: Some Phenomenological Predictions from Cognitive Neuroscience," first appeared in *Philosophical Psychology* 18, no. 5 (2005).

"On the Reality (and Diversity) of Objective Colors: How Color-Qualia Space Is a Map of Reflectance-Profile Space," is currently in press at *Philosophy of Science* (2006).

"Into the Brain: Where Philosophy Should Go from Here," first appeared in *Topoi* 25 (2006): 29–32, a special issue on the future of philosophy.

1

Catching Consciousness in a Recurrent Net

Dan Dennett is a closet Hegelian. I say this not in criticism, but in praise, and hereby own to the same affliction. More specifically, Dennett is convinced that human cognitive life is the scene or arena of a swiftly unfolding evolutionary process, an essentially cultural process above and distinct from the familiar and much slower process of biological evolution. This superadded Hegelian adventure is a matter of a certain style of *conceptual* activity; it involves an endless contest between an evergreen variety of conceptual *alternatives*; and it displays, at least occasionally, a welcome *progress* in our conceptual sophistication, and in the social and technological practices that structure our lives.

With all of this, I agree, and will attempt to prove my fealty in due course. But my immediate focus is the peculiar *use* to which Dennett has tried to put his background Hegelianism in his provocative 1991 book, *Consciousness Explained.*[1] Specifically, I wish to address his peculiar account of the *kinematics and dynamics* of the Hegelian Unfolding that we both acknowledge. And I wish to query his novel *deployment* of that kinematics and dynamics in explanation of the focal phenomenon of his book: consciousness. To state my negative position immediately,

---

[1] (Boston: Little, Brown, 1991). I first addressed Dennett's account of consiousness in *The Engine of Reason, the Seat of the Soul: A Philosophical Journey into the Brain* (Cambridge, MA: MIT Press, 1995), 264–9. A subsequent two-paper symposium appears as S. Densmore and D. Dennett, "The Virtues of Virtual Machines," and P. M. Churchland, "Densmore and Dennett on Virtual Machines and Consciousness," *Philosophy and Phenomenological Research* 59, no. 3 (Sept., 1999): 747–67. This essay is my most recent contribution to our ongoing debate, but Dennett has a worthy reply to it in a recent collection of essays edited by B. L. Keeley, *Paul Churchland* (New York: Cambridge University Press, 2005), 193–209.

I am unconvinced by his declared account of the background process of human conceptual evolution and development – specifically, the Dawkinsean account of rough gene-analogs called "memes" competing for dominance of human cognitive activity.[2] And I am even less convinced by Dennett's attempt to capture the emergence of a peculiarly human consciousness in terms of our brains' having internalized a specific complex *example* of such a "meme," namely, the serial, discursive style of cognitive processing typically displayed in a von Neumann computing machine.

My opening task, then, is critical. I think Dennett is wrong to see human consciousness as the result of a unique form of "software" that began running on the existing hardware of human brains some ten, or fifty, or a hundred thousand years ago. He is importantly wrong about the character of that background software process in the first place, and he is wrong again to see consciousness itself as the isolated result of its "installation" in the human brain. Instead, as I shall argue, the phenomenon of consciousness is the result of the brain's basic *hardware* structures, structures that are widely shared throughout the animal kingdom, structures that produce consciousness in meme-free and von Neumann–innocent animals just as surely and just as vividly as they produce consciousness in us. As my title indicates, I think the key to understanding the peculiar weave of cognitive phenomena gathered under the term "consciousness" lies in understanding the dynamical properties of biological neural networks with a highly *recurrent* physical architecture – an architecture that represents a widely shared hardware feature of animal brains generally, rather than a unique software feature of human brains in particular.

On the other hand, Dennett and I share membership in a small minority of theorists on the topic of consciousness, a small minority even among materialists. Specifically, we both seek an explanation of consciousness in the *dynamical* signature of a conscious creature's cognitive activities rather than in the peculiar character or subject matter of the *contents* of that creature's cognitive states. Dennett may seek it in the dynamical features of a "virtual" von Neumann machine, and I may seek it in the dynamical features of a massively recurrent neural network, but we are both working the "dynamical profile" side of the street, in substantial isolation from the rest of the profession.

Accordingly, in the second half of this paper I intend to defend Dennett in this dynamical tilt, and to criticize the more popular content-focused

---

[2] As outlined in M. S. Dawkins, *The Selfish Gene* (Oxford: Oxford University Press, 1976), and Dawkins, *The Extended Phenotype* (San Francisco: Freeman, 1982).

alternative accounts of consciousness, as advanced by most philosophers and even by some neuroscientists. And in the end, I hope to convince both Dennett and the reader that the hardware-focused recurrent-network story offers the most fertile and welcoming reductive home for the relatively unusual dynamical-profile approach to consciousness that Dennett and I share.

I. Epistemology: Naturalized and Evolutionary

Attempts to reconstruct the canonical problems of epistemology within an explicitly evolutionary framework have a long and vigorous history. Restricting ourselves to the twentieth century, we find, in 1934, Karl Popper already touting experimental falsification as the selectionist mechanism within his expressly evolutionary account of scientific growth, an account articulated in several subsequent books and papers.[3] In 1950, Jean Piaget published a broader and much more naturalistic vision of information-bearing structures in a three-volume work assimilating biological and intellectual evolution.[4] Thomas Kuhn's 1962 classic[5] painted an overtly antilogicist and anticonvergent portrait of our scientific development, and proposed instead a radiative process by which different cognitive paradigms would evolve toward successful domination of a wide variety of cognitive niches. In 1970, and partly in response to Kuhn, Imre Lakatos[6] published a generally Popperian but much more detailed account of the dynamics of intellectual evolution, one more faithful to the logical, sociological, and historical facts of our own scientific history. In 1972, Stephen Toulmin[7] was pushing a biologized version of Hegel, and by 1974 Donald Campbell[8] had articulated a deliberately Darwinian account of the blind generation and selective retention of scientific theories over historical time.

---

[3] *Logik der Forschung* (Wien, 1934). Published in English as *The Logic of Scientific Discovery* (London: Hutchison, 1980). See also Poppers's *locus classicus* essay, "Conjectures and Refutations," in his *Conjectures and Refutations* (London: Routledge, 1972). See also Popper, *Objective Knowledge: An Evolutionary Approach* (Oxford: Oxford University Press, 1979).

[4] *Introduction a l'epistemologie genetique*, 3 vols. (Paris: Presses Universitaires de France, 1950). See also Piaget, *Insights and Illusions of Philosophy* (New York: Meridian Books, 1965), and Piaget, *Genetic Epistemology* (New York: Columbia University Press 1970).

[5] *The Structure of Scientific Revolutions* (Chicago: University of Chicago Press, 1962).

[6] "Falsification and the Methodology of Scientific Research Programs," in I. Lakatos and A. Musgrave, eds., *Criticism and the Growth of Knowledge* (Cambridge: Cambridge University Press, 1970).

[7] S. Toulmin, *Human Understanding* (Princeton, NJ: Princeton University Press, 1972).

[8] "Evolutionary Epistemology," in *The Philosophy of Karl Popper*, P. A. Schilpp, ed. (La Salle, IL: The Open Court, 1974).

From 1975 on, the literature becomes too voluminous to summarize easily, but it includes Richard Dawkins's specific views on memes, as scouted briefly in *The Selfish Gene* (1976) and more extensively in *The Extended Phenotype* (1982). In some respects, Dawkins's peculiar take on human intellectual history is decidedly better than the take of many others in this tradition – most important, his feel for both genetic theory and biological reality is much better than that of his precursors. In other respects, it is rather poorer – comparatively speaking, and once again by the standards of the tradition at issue. Dawkins is an epistemological naïf, and his feel for our actual scientific/conceptual history is rudimentary. But he had the wit, over most of his colleagues, to escape the biologically naïve construal of theories-as-*genotypes* or theories-as-*phenotypes* that attracted so many other writers. Despite a superficial appeal, both of these analogies are deeply strained and ultimately infertile, both as extensions of existing biological theory and as explanatory contributions to existing epistemological theory.[9] Dawkins embraces, instead, and despite my opening characterization, a theories-as-*viruses* analogy, wherein the human brain serves as a host for competing invaders, invaders that can replicate by subsequently invading as-yet uninfected brains.

While an improvement in several respects, this analogy seems stretched and problematic still, at least to these eyes. An individual virus is an individual physical thing, locatable in space and time. An individual theory is no such thing. And even its individual "tokens" – as they may be severally embodied in the distinct brains they have "invaded" – are, at best, abstract *patterns* of some kind imposed upon preexisting physical structures within the brain, not physical *things* bent on making further physical things with a common physical structure.

Further, a theory has no internal mechanism that effects a literal self-replication when it finds itself in a fertile environment, as a virus has when it injects its own genetic material into the interior of a successfully hijacked cell. And my complaint here is not that the mechanisms of self-replication are different across the two cases. It is that there *is no* such mechanism for theory tokens. If they can be seen as "replicating" at all, it must be by some wholly different process. This is further reflected in the fact that theory tokens do not replicate themselves *within* a given individual, as viruses most famously do. For example, you might have 10^6

---

[9] An insightful perspective on the relevant defects is found in C. A. Hooker, *Reason, Regulation, and Realism: Toward a Regulatory Systems Theory of Reason and Evolutionary Epistemology* (Albany, NY: SUNY Press, 1995), 36–42.

qualitatively identical rhinoviruses in your system at one time, all children of an original invader; but never more than one token of Einstein's theory of gravity.

Moreover, the brain is a medium selected precisely for its ability to assume, hold, and deploy the conceptual systems we call theories. Theories are not alien invaders bent on subverting the brain's resources to their own selfish "purposes." On the contrary, a theory is the brain's way of making sense of the world in which it lives, an activity that is its original and primary function. A bodily cell, by contrast, enjoys no such intimate relationship with the viruses that intrude upon its normal metabolic and reproductive activities. A mature cell that is completely free of viruses is just a normal, functioning cell. A mature brain that is completely free of theories or conceptual frameworks is an utterly dysfunctional system, barely a brain at all.

Furthermore, theories often – indeed, usually – take *years* of hard work and practice to grasp and internalize, precisely because there is no analog to the physical virus entering the body, pill-like or bullet-like, at a specific time and place. Instead, a vast reconfiguration of the brain's 10^{14} synaptic connections is necessary in order to imprint the relevant conceptual framework on the brain, a reconfiguration that often takes months or years to complete. Accordingly, the "replication story" needed, on the Dawkinsean view, must be nothing short of an entire theory of how the brain *learns*. No simple "cookie-cutter" story of replication will do for the dubious "replicants" at this abstract level. There are no zipper-like molecules to divide down the middle and then reconstitute themselves into two identical copies. Nor will literally repeating the theory, by voice or in print, to another human do the trick. Simply receiving, or even memorizing, a list of presented *sentences* (a statement of the theory) is not remotely adequate to successful acquisition of the conceptual framework to be replicated, as any unprepared student of classical physics learns when he or she desperately confronts the problem-set on the final examination, armed only with a crib sheet containing flawless copies of Newton's gravitation law and the three laws of motion. Knowing a theory is not just having a few lines of easily transferable syntax, as the student's inevitable failing grade attests.

The poverty of its "biological" credentials aside, the *explanatory payoff* for embracing this viruslike conception of theories is quite unremarkable in any case. The view brings with it no compelling account of where theories originate, how they are modified over time in response to experimental evidence, how competing theories are evaluated, how they guide our experimental and practical behaviors, how they fuel our technological economies, and how they count as representations of the world's hidden structure. In short, the analogy with viruses does not provide particularly illuminating answers, or any answers at all, to most of the questions that make up the problem-domain of epistemology and the philosophy of science.

What it does do is hold out the promise of a grand consilience – a conception of scientific activity that is folded into a larger and more powerful background conception of biological processes in general. This is, at least in prospect, an extremely *good* thing, and it more than accounts for the "aha!" feelings that most of us experience upon first contemplating such a view. But closer examination shows it to be a *false* consilience, based on a false analogy. Accordingly, we should not have much confidence in deploying it, as Dennett does, in hopes of illuminating either human cognitive development in general, or the development of human consciousness in particular.

Despite reaching a strictly negative conclusion here, not just about the theories-as-viruses analogy but about the entire evolutionary tradition in recent epistemology, I must add that I still regard that tradition as healthy, welcome, and salutary, for it seeks a worthy sort of consilience, and it serves as a vital foil against the deeply sclerotic logicist tradition of the logical empiricists. Moreover, I share the background conviction of most people working in the newer tradition – namely, that in the end a proper account of human scientific knowledge must somehow be a proper part of a general theory of biological systems and biological development. However, I have quite different expectations about how that integration should proceed. They are the focus of a book in progress, but the present occasion is focused on consciousness, so I must leave their articulation for another time. In the meantime, I recommend C. A. Hooker's "nested hierarchy of regulatory mechanisms" attempt – to locate scientific activity within the embrace of biological phenomena at large – as the most promising account in the literature.[10] We now return to Dennett.

II. The Brain as Host for the von Neumann Meme

If the human brain *were* a von Neumann machine (hereafter, vN machine) – literally, rather than figuratively or virtually – then the virus

---

[10] Hooker, *Reason, Regulation, and Realism*, 36–42. For a review of Hooker's book and its positive thesis, see P. M. Churchland, "Review of *Reason, Regulation, and Realism*," *Philosophy and Phenomenological Research* 58, no. 4 (1999): 541–4.

analogy just rejected would have substantially more point. We do speak of, and bend resources to avoid, "computer viruses," and the objections voiced earlier, concerning theories and the brain, are mostly irrelevant if the virus analogy is directed instead at programs loaded in a computer. A program *is* just a package of syntax; a program *can* download in seconds; a program *can* contain a self-copying subroutine; and a program *can* fill a hard drive with monotonous copies of itself, whether or not it ever succeeds in infecting a second machine.

But the brains of animals and humans are most emphatically *not* vN machines. Their coding is not digital; their processing is not serial; they do not execute stored programs; and they have no random-access storage registers whatever. As fifty years of neuroscience and fifteen years of neuromodeling have taught us, a brain is a different kettle of fish entirely. That is why brains are so hopeless at certain tasks, such as multiplying two twenty-digit numbers in one's head, which task a computer does in a second. And that is why computers are so hopeless at certain other tasks, such as recognizing individual faces or understanding speech, which task a brain does in even less time.

We now know enough about both brains and vN computers to appreciate precisely why the brain does as well as it does, despite being made of components that are a million times slower than those of an electronic computer. Specifically, the brain is a massively parallel vector processor. Its background understanding of the world's general features (its conceptual framework) resides in the slowly acquired configuration of its 10^{14} synaptic connections. Its specific understanding of the local world here-and-now (its fleeting thoughts and perceptions) resides in the fleeting patterns or vectors of activation-levels across its 10^{11} neurons. And the character of those fleeting patterns is dictated by the learned matrix of synaptic connections that serve simultaneously to transform *peripheral* sensory activation vectors into well-informed *central* vectors, and ultimately into the well-orchestrated *motor* vectors that produce our bodily behavior.

Now Dennett knows all of this as well as anyone, and it poses a problem for him. It's a problem because, as discussed earlier, the virus analogy that he intends to exploit requires a vN computer for its plausibility. But the biological brain is not a vN computer. So Dennett postulates that, at some point in our past, the human brain managed to "reprogram" itself in such a fashion that its genetically endowed "hardware" came to "load" and "run" a peculiar piece of novel "software" – an invading virus or meme – such that the brain came to *be* a "virtual" von Neumann machine.

But wait a minute. We are here contemplating an explanation – of how the brain *came to be* a virtual vN machine – in terms that make clear and literal sense only if the brain was *already* a (literal) vN machine. But it wasn't. And so it couldn't become *any* new "virtual" machine – and a fortiori not a virtual vN machine – in the literal fashion described. Dennett must have some related but metaphorical use in mind for the expressions "program," "software," "hardware," "load," and "run." And, as we shall see, for "virtual" and "vN machine" as well.

As indeed he does. Dennett knows that brains are plastic in their configurations of synaptic connections, and he knows that changing those configurations produces changes in the way the brain processes information. He is postulating that, at some point in the past, at least one human brain lucked/stumbled into a global configuration of synaptic connections that embodied an importantly new style of information processing, a style that involved, at least occasionally, the sequential, temporally structured, rule-respecting kinds of activities seen in a typical vN machine.

Let us look into this possibility. What is the actual potential of a massively parallel vector-processing machine to "simulate" a vN machine? For a purely feedforward network (Figure 1.1 *a*), it is zero, because such a network cannot execute the temporally *recursive* procedures essential to a program-executing vN machine. To surmount this trivial limitation, we need to step up to networks with a *recurrent* architecture (Figure 1.1 *b*), for as is well known, this is what permits any neural network to deal with structures in time.

Artificial recurrent networks have indeed been trained up to execute successfully the kinds of explicitly recursive procedures involved in, for example, adding individual pairs of *n*-digit numbers,[11] and distinguishing grammatical from ungrammatical sentences in a (highly simplified) productive language.[12]

But are these suitably trained networks thus "virtual" adders and "virtual" parsers? No. They are *literal* adders and parsers. The language of "virtual machines" is not strictly appropriate here, because these are *not* cases of a special purpose "software machine" running, qua program, on a vN-style universal Turing machine.

More generally, the idea that a machine, any machine, might be programmed to "simulate" a vN machine in particular makes the mistake of treating *vN machine* as if it were itself a *special*-purpose piece of software,

---

[11] G. W. Cottrell, and F. Tsung, "Learning Simple Arithmetic Procedures," *Connection Science* 5, no. 1 (1993): 37–58.

[12] J. L. Elman, "Grammatical Structure and Distributed Representations," in S. Davis, ed., *Connectionism: Theory and Practice,* vol. 3 in the series Vancouver Studies in Cognitive Science (Oxford: Oxford University Press, 1992), 138–94.

rather than what it is, namely, an entirely *general*-purpose organization of *hardware*. In sum, the brain is not a machine that is capable of "downloading software" in the first place, and a vN machine is not a piece of "software" fit for downloading in any case.

Accordingly, I cannot find a workable interpretation of Dennett's proposal here that is both nonmetaphorical and true. Dennett seems to be trying to both eat his cake (the brain becomes a vN machine by downloading some software) and have it too (the brain is not a vN machine to begin with). And these complaints are additional to and independent of the complaints of the preceding section, to the effect that Dawkins's virus analogy for cultural acquisitions such as theories, songs, and practices is a false and explanatorily sterile analogy to begin with.

There is an irony here. The fact is, if we do look to recurrent neural networks – which brains most assuredly are – in order to purchase something like the functional properties of a vN machine, we no longer *need* to "download" any epigenetically supplied meme or program, because the sheer hardware configuration of a recurrent network already delivers the desired capacity for recognizing, manipulating, and generating serial structures in time, right out of the box. Those characteristic recurrent pathways are the very computational resource that allows us to recognize a puppy's gait, a familiar tune, a complex sentence, and a mathematical proof. Which *particular* temporal structures come to dominate a network's cognitive life will be a function of which causal processes are perceptually encountered during its learning phase. But the need for a virtual vN machine, in order to achieve this broader family of cognitive ends, has now been lifted. The brain doesn't need to import the "software" Dennett contrives for it: its existing "hardware" is already equal to the cognitive tasks that he (rightly) deems important.

This fact moves me to try to reconstruct a vaguely Dennettian account of consciousness using the very real resources of a recurrent physical architecture, rather than the strained and figurative resources of a virtual vN machine. And this brings me to the dynamical-profile approach cited at the outset of this paper. But first I must motivate its pursuit by evoking and dismantling its principal explanatory adversary, the content-focused approach.

III. Consciousness as Self-Representation: Some Problems

One strategy for trying to understand consciousness is to see it as a species of *representation*, a species distinguished by its peculiar *contents*, specifically, the current states or activities of the *self*, that is, the current states or activities of the very biological-cum-cognitive system engaged in such representation. Consciousness, on this view, is essentially a matter of self-perception or self-representation. Thus, one is conscious when, for example, one's cognitive system represents stress or damage to some part of one's body (pain), when it represents one's empty stomach (hunger), when it represents the postural configuration of one's body (hands folded in front of one), when it represents one's high-level cognitive state ("I believe Budapest is in Hungary"), or when it represents one's relation to an external object ("I'm about to be hit by an incoming snowball").

Kant's doctrine of inner sense in *The Critique of Pure Reason* is the classic (and highly a priori) instance of this approach, and Antonio Damasio's book *The Feeling of What Happens*[13] provides a modern (and neurologically grounded) instance of the same general strategy. While I have some sympathy for this approach to consciousness – I have defended it myself in *Matter and Consciousness*[14] – this chapter is aimed at overturning it and replacing it with a specific alternative. Let me begin by voicing the central worries – to which all parties must be sensitive – that cloud the self-representation approach to consciousness.

There are two major weaknesses in the approach. The first is that it fails, at least on all outstanding versions, to give a clear and adequate account of the inescapable distinction between those of our self-representations that are conscious and those that are not. The nervous system has a great many subsystems that continuously monitor a wide variety of visceral, hormonal, thermal, metabolic, and other regulatory activities of the biological organism. These are representations of the self, if anything is, but they are only occasionally a part of our consciousness, and some of them are *permanently* beneath the level of conscious awareness.

One might try to avoid this difficulty by stipulating that the self-representations that constitute the domain of consciousness must be representations of the states and activities of the brain and nervous system proper, rather than of the body in general. But this proposal has three daughter difficulties. Prima facie, the stipulation would *exclude* far too much, for hunger, pain, and other plainly conscious somatosensory sensations are clearly representations of various aspects of the body, not the brain. Less obviously, but equally problematic, it would falsely *include* the

[13] (New York: Harcourt, 1999).
[14] Rev. ed. (Cambridge, MA: MIT Press, 1986), 73–5, 119–20, 179–80.

enormous variety of brain activities that constitute ongoing and systematic representations of other aspects of the brain itself – indeed, these are the bulk of them – but which never make it into the spotlight of consciousness. We must be mindful, that is, that most of the brain's representational activities are self-directed and lie well below the level of conscious awareness. Finally, the proposed stipulation would wrongly *exclude* from consciousness the brain's unfolding representations of the world beyond the body, such as our visual awareness of the objects at arm's length and our auditory awareness of the whistling kettle. One might try to insist that, strictly speaking, it is only our visual and auditory *sensations* of which we are directly conscious – external objects being only indirect and secondary objects of awareness – but this move is false to the facts of both human cognitive development and human phenomenology, and it leads us down the path of classical sense-datum theory, whose barrenness has long been apparent.

A special *subject matter*, then, seems not to be the essential feature that distinguishes conscious representations from all others. To the contrary, it would seem that a conscious representation could have any content or subject matter at all. The proposal under discussion would seem to be confusing *self*-consciousness with consciousness in general. The former is highly interesting, to be sure, but it is the latter that is our current explanatory target.

The self-representation view has a second major failing, which emerges as follows. Consider a creature, such as you or me, who has a battery of distinct sensory modalities – a visual system, an auditory system, an olfactory system – for constructing representations of various aspects of the physical world. And suppose further that, as cognitive theorists, we have some substantial understanding of how those several modalities actually work, as devices for monitoring aspects of external reality and coding those aspects internally. And yet we remain mystified about what makes the representations in which they trade *conscious* representations. We remain mystified, that is, at what distinguishes the conscious states of one's visual system from the equally representational but utterly unconscious representational states of a voltmeter, an audio tape recorder, or a video camera. Now, if our general problem is thus to try to understand how *any* representational modality ascends to the level of conscious representations, then proposing a proprietary representational modality whose job it is to monitor phenomena *inside* the skin, rather than outside the skin, is a blatant case of *repeating* our problem, not of solving it. Our original problem attends the inward-looking modality no less than the various outward-looking modalities with which we began, and adding the inward modality does nothing obvious to transform the outward ones in any case. Once again, leaning on the *content* of the representations at issue – on the *focus, target,* or *subject matter* of the epistemic modality in question – fails to provide the explanatory factors that we seek. We need to look elsewhere.

IV. The Dynamical-Profile Approach

We need to look, I suggest, at the peculiar *activities* in which some of our representations participate, and at the special computational context required for those activities to take place. I here advert, for example, to the brain's capacity (1) to focus attention on some aspect or subset of its teeming polymodal sensory inputs, (2) to try out different conceptual interpretations of that selected subset, and (3) to hold the results of that selective/interpretive activity in short-term memory for long enough (4) to update a coherent representational "narrative" of the world-un-folding-in-time, a narrative thus fit for possible selection and imprinting in long-term memory.

Any cognitive representation that figures in the dynamical/computational profile just outlined is a recognizable candidate for, and a presumptive instance of, the class of *conscious* representations. We may wish to demand still more of such candidates than merely meeting these quick four conditions, but even these four specify a dynamical or functional profile recognizable as typical of conscious representations. Notice also that this profile makes no reference to the specific *content,* either semantic or qualitative, of the representation that meets it, reflecting the fact, agreed to in the last section, that a conscious representation could have any content whatever.

Appealing to notions such as attention, interpretation, and short-term memory may seem, however, to be just helping oneself to a handful of notions that are as opaque or problematic as the notion of consciousness itself, unless we can provide independent explanations of these dynamical notions in neuronal terms. In fact, that is precisely what the dynamical properties of recurrent neural networks allow us to do, and more besides, as I shall now try to show.

The consensus concerning information processing in artificial neural networks is that their training history slowly produces a *sculpted space* of possible representations (= possible activation patterns) at any given layer or population of neurons (such as the middle layer of the network in Figure 1.1*a*). Such networks, trained to discriminate or recognize

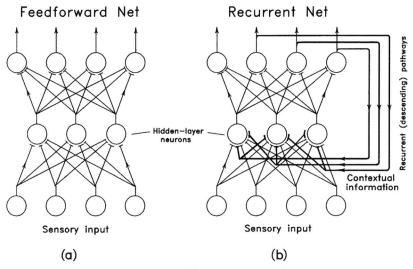

FIGURE 1.1. Elementary networks

instances of some range of categories, $c_1, \ldots, c_2$, slowly acquire a corresponding family of "attractors" or "prototype wells" variously located within the space of possible activation patterns. That sculpted space *is* the conceptual framework of that layer of neurons. Diverse sensory-layer instances of those learned perceptual categories produce activation patterns within, or close to, one or another of these "preferred" prototype regions within the activation space of the second layer of neurons.

Purely feedforward networks can achieve quite astonishing levels of discriminatory skill, but beyond a welcome tendency to "fill in" or "complete" degraded or partial perceptual instances of the categories to which they have been trained,[15] they are rather dull and predictable fellows. However, if we add recurrent or descending pathways to the basic feedforward architecture, as in Figure 1.1*b*, we lift ourselves into a new universe of functional and dynamical possibilities.

For example, information from the higher levels of any network – information that is the result of somewhat earlier information processing by the network – can be entered as a supplementary "context fixer" at the second layer of the network. This information can and does serve to "prime" or "prejudice" that neuronal population's collective activity in the direction of one or another of its learned perceptual categories.

---

[15] See pp. 45–6 and 107–14 of Churchland, *The Engine of Reason, the Seat of the Soul,* for a more detailed discussion of this intriguing feature of feedforward network activity.

The network's cognitive "attention" is now preferentially focused on one of its learned categories at the expense of the others. That is to say, the probability that that focal prototype category will be activated, given any arbitrary sensory input, has been temporarily raised, relative to all of its categorical alternatives.

Such an attentional focus is also movable, from one learned category to another, as a function of the network's unfolding activation patterns or "frame of mind" at its higher neuronal layers. Such a network has an ongoing *control* of its topical selections from, and its conceptual interpretations of, its unfolding perceptual inputs. In particular, such a network can bring to bear, now in a selective way, the general background knowledge embodied more or less permanently in the configuration of its myriad synaptic connections.

A recurrent architecture also provides the network with a grasp of *temporal* structure as well as of spatial structures. A feedforward network gives an invariant, one-shot response to any frozen "snapshot" pattern entered at its sensory layer. But a recurrent network can represent the changing perceptual world with a continuous *sequence* of activation patterns at its second layer, as opposed to a single, fixed pattern. Indeed, what recurrent networks typically become trained to recognize are temporally structured *causal sequences*, such as the undulating pattern of a swimming fish, the trajectory of a bouncing ball, the loping gait of a running predator, or the grammatical structure of an uttered sentence. These phenomena are represented, at the second layer, not by a prototypical *point* in its sculpted activation space (as in a feedforward network), but by a prototypical *trajectory* within that space. Thus emerges a temporally structured "narrative" of the world-unfolding-in-time.

The recurrent pathways also bestow on the network a welcome form of short-term memory, one that is both topic-sensitive and has a variable decay time. For the second layer is in continuous receipt of a selectively processed "digest" of its own activity some t milliseconds ago, where t is the time it takes for an axonal message to travel up to the third layer and then back down again to the middle layer. Certain salient features of the middle-layer activation patterns, therefore, may survive many cycles of network activity, as a temporarily stable "limit cycle," before being displaced by some other limit cycle focused on some other perceptual category.

Since the network's behavior is now a continuous function of both its current perceptual inputs and its current dynamical (i.e., activational) state, we are looking at a genuine dynamical system with the capacity to display behaviors that are strictly unpredictable, short of our possessing infinitely accurate information about all of the interacting variables. That is to say, the system's future behavior will often be reliably predictable for very short distances into the future, such as a few seconds. And the gross outlines of some of its future behaviors may be reliably projected over periods of a day or a week (such as falling asleep each night or eating meals fairly regularly). But in between these two extremes, reliable prediction becomes utterly impossible. In general, the system is too mercurial to permit the prediction of absolutely specific behaviors at any point in the nonimmediate future. Thus emerges the spontaneity we expect of, and prize in, a normal stream of conscious cognitive activity.

Such spontaneity is a direct reflection of the operation of the recurrent pathways at issue, which operation yields another important feature of this architectural addition. With active descending pathways, input from the sensory layer is no longer necessary for the continued activity of the network. The information arriving at the middle layer by way of the descending pathways is entirely sufficient to keep that population of neurons humming away in representational activity, privately exploring the vast landscape of activational possibilities that make up its acquired activation space. Thus is day-dreaming made possible, and night-dreaming, too, for that matter, despite the absence of concurrent perceptual stimulation. Accordingly, and on the view proposed, the dynamical behaviors characteristic of consciousness do not require perceptual inputs at all. Evidently our unfolding perceptual inputs *regulate* those dynamical behaviors profoundly, unless one happens to be insane, but perceptual inputs are not strictly necessary for consciousness.

It is further tempting to see the selective *deactivation* of those recurrent pathways – leaving only the residual feedforward pathways on duty – as the key to producing so-called delta (i.e., deep or nondreaming) sleep. For in such a selectively deactivated condition, one's attention shuts down, one's short-term memory is deactivated, and one ceases entirely to control or modulate one's own cognitive activities. Functioning recurrent pathways are utterly essential to all of these things. The feed*forward* pathways presumably remain functional even when one is in deep sleep, because certain special perceptual inputs – such as an infant's voice or a scratching at the bedroom window – can be recognized and serve quickly to awaken one, even if those perceptual stimuli are quite faint. This is a simple job that even a feedforward network can do. Even an unconscious creature needs an alarm system to pick up on a small class of highly special perceptual inputs, and the residual feedforward pathways provide it.

But when morning breaks, the recurrent pathways come back on duty, and the peculiar dynamical profile of cognitive activities just detailed gets resurrected. One regains consciousness.

I will leave further exploration of these matters to another time, when I can better tie the story to the actual microanatomy of the brain.[16] The reader now has some sense of how some central features of consciousness might be explained in terms of the dynamical properties of neural networks having a recurrent architecture. I close by returning to Dennett, and I begin by remarking that, details aside, the functional or molar-level portrait of consciousness embodied in his multiple-drafts and fleeting-moments-of-fame metaphors is indeed another instance of what I have here been calling the dynamical-profile approach to understanding consciousness. But Dennett painted his portrait first, so it is appropriate for me to ask if *I* may belatedly come on board. I hope to be found a worthy cognitive ally in these matters. Even so, I present myself to him with a list of needed reforms. The virtual von Neumann machine and all the metaphors associated with it have to go. They lead us away from the shared truth at issue, not toward it.

At one point in his book, Dennett himself registers an important doubt concerning the explanatory payoff of the virtual vN machine story.

> But still (I am sure you want to object): all this has little or nothing to do with consciousness! After all, a von Neumann machine is entirely unconscious; why should implementing it – or something like it: a Joycean machine – be any more conscious? I do have an answer: The von Neumann machine, by being wired up from the outset that way, with maximally efficient informational links, didn't have to become the object of its own elaborate perceptual systems. The workings of the Joycean machine, on the other hand, are just as "visible" and "audible" to it as any of the things in the external world that it is designed to perceive – for the simple reason that they have much of the same perceptual machinery focused on them.[17]

Dennett's answer here is strictly correct, but it doesn't count as an *explanation* of why our Joycean/virtual-vN machine rises to consciousness while the real vN machine does not. It fails because it is an instance of the "self-perception" approach dismantled earlier in Section III. An inward-looking perceptual modality is just as problematic, where consciousness is concerned, as is any outward-looking perceptual modality.

---

[16] A first attempt appears in Churchland, *The Engine of Reason, the Seat of the Soul,* pp. 208–26. That discussion also locates the explanation of consciousness in particular within the context of intertheoretic reductions in general.

[17] Dennett, *Consciousness Explained,* 225–6.

The complaint here addressed by Dennett is a telling one, but Dennett's answer won't stand scrutiny. It represents an uncharacteristic lapse from his "dynamical-profile" story in any case.

The Dawkinsean meme story has to go also, and with it goes the idea that humans – that is, animals genetically and neuroanatomically identical with modern humans – developed or stumbled upon consciousness as a purely cultural addition to our native cognitive machinery. On the contrary, we have been conscious creatures for as long as we have possessed our current neural architecture. Further, the contrast between human and animal consciousness has to go as well, for nonhuman animals *share* with us the recurrent neuronal architecture at issue. Accordingly, conscious cognition has presumably been around on this planet for at least fifty million years, rather than for the several tens of thousands of years guessed by Dennett.

I do not hesitate to concede to Dennett that cultural evolution – the Hegelian Unfolding that we both celebrate – has succeeded in "raising" human consciousness profoundly. It has raised it in the sense that the *contents* of human consciousness – especially in our intellectual, political, artistic, scientific, and technological elites – have been changed dramatically. Old conceptual frameworks, in all of the domains listed, have been discarded wholesale in favor of new frameworks, frameworks that underwrite new forms of human perception and new forms of human activity. Nor do I think we are remotely done yet, in this business of cognitive self-reconstruction. Readers of my 1979 book[18] will not be surprised to hear me suggesting still that the great bulk and most dramatic increments of consciousness-raising lie in our future, not in our past.

But raising the contents of our consciousness is one thing – and so far, largely a cultural thing. *Creating* consciousness in the first place, by contrast, was a firmly *neurobiological* thing, and that must have happened a very long time ago. For the dynamical cognitive profile that constitutes consciousness has been the possession of terrestrial creatures since at least the early Jurassic. James Joyce and John von Neumann were simply not needed.

---

[18] *Scientific Realism and the Plasticity of Mind* (Cambridge: Cambridge University Press, 1979). On this point, see especially chaps. 2 and 3.

2

Functionalism at Forty

A Critical Retrospective

For those of us who were undergraduates in the 1960s, functionalism in the philosophy of mind was one of the triumphs of the new analytic philosophy. It was a breath of theoretical fresh air, a framework for conceptual clarity and computational rigor, and a shining manifesto for the possibility of artificial intelligence. Those who had been logical behaviorists rightly embraced it as the natural and more penetrating heir to their own deeply troubled views. Those who had been identity theorists embraced it as a more liberal but still agreeably robust form of scientific materialism. Those many who hoped to account for cognition in broadly computational terms found, in functionalism, a natural philosophical home. Even the dualists who refused to embrace it had to give grudging approval for its strictly antireductionist stance. It had something for everyone. Small wonder that it became, and has largely remained, the dominant position in the philosophy of mind, and, perhaps more importantly, in cognitive psychology and classical AI research as well.

Whether it still deserves that position – indeed, whether it ever did – is the principal subject of this essay. The legacy of functionalism, now visible to everyone after forty years of philosophical and scientific research, has not been entirely positive. But let us postpone criticism for a moment, and remind ourselves of the central claims that captured so many imaginations.

I. The Central Claims of Classical Functionalism

1. What unites all cognitive creatures is not that they share the same computational mechanisms (their 'hardware'). What unites them

is that (plus or minus some individual defects or acquired special skills) they are all computing the same, or some part of the same, abstract ⟨⟨sensory input, prior state⟩, ⟨motor output, subsequent state⟩⟩ *function.*[1]

2. The central job of cognitive psychology is to *identify* this abstract function that we are all (more or less) computing.

3. The central job of AI research is to create *novel physical realizations* of salient parts of, and ultimately all of, the abstract function we are all (more or less) computing.

4. Folk psychology – our commonsense conception of the causal structure of cognitive activity – already embodies a crude and partial representation of the function we are all (more or less) computing.

5. The reduction of folk psychology (indeed, any psychology) to the neuroscience of human brains is twice impossible, because:
   a. the relevant function is computable in a potentially infinite variety of ways, not just in the way that humans happen to do it, and
   b. such diverse computational procedures are in any case realizable in a potential infinity of distinct physical substrates, not just in the specifically human biological substrate.

   Accordingly, to reduce the categories of folk psychology to the idiosyncratic procedures and mechanisms of specifically *human* brain activity would be to *exclude*, from the domain of genuine cognitive agents, the endless variety of other realizations of the characteristic function (see point 1) that we are all computing. The kind-terms of psychology must thus be functionally rather than naturalistically or reductively defined.

6. Empirical research into the microstructure and microactivities of human and animal brains is entirely legitimate (for certainly we do wish to know how the sought-after function is *realized* in our own idiosyncratic case). But it is a very poor research strategy for recovering the global function itself, whose structure will be more

---

[1] Just to remind, a function is a set of input–output pairs, such that for each possible input, there is assigned a unique output. Such sets can have infinitely many input–output pairs, and the relations between the inputs and outputs can display extraordinary levels of complexity. The characterization proposed in point 1 is thus in no sense demeaning to cognitive creatures. It requires only that the relevant function be computable, i.e., that the proper output for any given input can be recursively generated by a finite system, such as a brain, in a finite time.

instructively revealed in the situated molar-level behavior of the entire creature.

7. Points 5 and 6 jointly require us to respect and defend the *methodological autonomy* of cognitive psychology, relative to such lower-level sciences as brain anatomy, brain physiology, and bio-chemistry. Cognitive psychology is picking up on its own laws at its own level of physical complexity.

Thus the familiar and collectively compelling elements of a highly influential philosophical position. Perhaps astonishingly, the position is decisively mistaken in all seven of the elements just listed. Or so, at least, I shall argue in what follows.

II. Some Unexpected Lessons from Neurobiology

The classical or 'program-writing' research tradition in AI was one highly promising expression of the functionalist view just outlined. But by the early 1980s, that research program had hit the wall with an audible thud. Despite the development of central processing units with increasingly fabulous clock speeds (even desktop machines now top 10^9 hertz), despite ever-expanding memory capacities (even desktop machines now boast over 10^{10} bytes), despite blistering internal signal conduction velocities (close to the speed of light), and despite the continuing a priori assurance (grounded in the Church-Turing thesis) that a universal Turing machine could, in principle, compute any computable function whatever, programmed computers in fact performed very poorly relative to their biological counterparts, at least on a wide variety of typical cognitive tasks.

The problem was not that there was any well-defined class of cognitive tasks that programmed digital computers proved utterly unable to even begin to simulate. The problem was rather that equal increments of progress toward more realistic cognitive simulations proved to require the commitment of exponentially increasing resources in memory capacity, computational speed, and program complexity. Moreover, even when sufficient memory capacity was made available to cover all of the empirical contingencies that real cognition is prepared to encounter, a principled way of retrieving, from that vast store, all and only the *currently relevant* information proved entirely elusive. As the memories were made larger, the retrieval problem got worse. Accordingly, as the computers' actual cognitive performance approached the levels displayed by biological brains (and in many cases they did), the time taken for the machines to produce the desired performance expanded to ridiculous lengths. A programmed machine took minutes or hours to do what a biological brain could do in a fraction of a second.

At the time, this was deeply puzzling, because no process in the brain had a 'clock frequency' higher than perhaps 100 hertz, and because typical signal conduction velocities within the brain are no greater than the speed of a human bicycle rider: perhaps 10 m/sec. In the respects at issue, this puts the biological brain at an enormous disadvantage: $\approx 10^2$ Hz vs. $\approx 10^9$ Hz in the first dimension of performance, and ≈ 10 m/sec vs. $\approx 10^8$ m/sec in the second. All told then, the computer should have a computational speed advantage of roughly $10^7 \times 10^7 = 10^{14}$, or fourteen orders of magnitude. And yet, as we now say, shaking our heads in amazement, the presumptive tortoise (the biological brain) easily outruns the presumptive hare (the electronic digital computer), at least on a wide variety of typical cognitive tasks.

The explanation of the human brain's impressively high performance, despite the very real handicaps mentioned, is no longer a matter of controversy. The brains of terrestrial creatures all deploy a computational strategy quite different from that deployed in a standard serial-processing, digital-coding, electronic computer. That strategy allows them to do a clever end run around their time-related handicaps. Specifically, the biological brain is a massively *parallel* piece of computational machinery: it performs trillions of individual computational transformations – within the 10^{14} individual microscopic *synaptic connections* distributed throughout its volume – *simultaneously* and *all at once*. And it can repeat such feats of computation at least ten and perhaps a hundred times per second. The presumptive deficit of fourteen orders of magnitude scouted earlier is thus made good in one fell swoop. And the brain is left with a modest computational advantage of its own concerning the number of basic computational operations performed per second: perhaps one or two orders of magnitude over current electronic machines.

Moreover, this massively parallel, distributed processing (or "PDP," as it has come to be called) provides a built-in solution to classical AI's chronic problem of how to access, in real time and from the totality of one's vast memory store, all and only the informational elements that are relevant to one's current computational problem. The fact is, the acquired strengths or 'weights' of the brain's 10^{14} synaptic connections collectively embody *all* of the acquired wisdom and acquired skills that the creature commands. (Learning, at least in its most basic form, *consists in* the progressive modification of those myriad connections.) But those

100 trillion synaptic connections are also the brain's basic *computational* elements. Each time a large cadre of synaptic connections effects a transformation of an incoming representation into an output representation at the receiving population of neurons, *every synapse in that entire cadre* has a hand in shaping that computational transformation, and each makes its tiny contribution simultaneously with all of the others.

Accordingly, it is not just the brain's computational behavior that is massively parallel. Its *access to memory* is also a massively parallel affair. Indeed, these are no longer distinct processes, as they are in a digital computer with a classical von Neumann architecture. In the biological brain, to engage in any computational transformation simply *is* to deploy whatever knowledge the brain has accumulated. Thus, the classical Frame Problem[2] for artificial intelligence simply evaporates, as does the Inductive Logician's Problem of the global sensitivity (to background knowledge) of any abductive inference,[3] which is easily the most common form of inference that any creature ever performs.

These welcome developments concerning the general nature of information processing in humans and animals were humbling for the ambitions of classical AI not because those ambitions were revealed to be unachievable. On the contrary, artificial intelligence now looks more achievable than ever. Rather, these decisively illuminating developments were humbling because they were the result of empirical and theoretical research within two *lower-level* sciences, neuroanatomy and neurophysiology, whose contributions to cognitive psychology and AI were widely and officially expected to be minimal at best, and procrustean at worst. (See again points 5), 6), and 7).) But those often-derided 'engineering details' turned out to be decisively relevant to understanding how a plodding biological brain could keep up with an electronic machine in the first place. And they proved equally decisive for understanding how the brain could routinely solve a vital *cognitive* problem – the real-time selective deployment of relevant information – that the programmed serial machines were quite unable to solve. Cognitive psychology, it began to emerge,

---

[2] D. C. Dennett, "Cognitive Wheels: The Frame Problem in Artificial Intelligence," in C. Hookway, ed., *Minds, Machines, and Evolution* (Cambridge: Cambridge University Press, 1984).

[3] For a recent summary, see J. A. Fodor, "The Mind Doesn't Work That Way" (Cambridge, MA: MIT Press, 2000). Also, P. M. Churchland, "Inner Spaces and Outer Spaces: The New Epistemology" (in preparation), chap. 2.

was not so 'methodologically autonomous' as the functionalists had advertised.

III. Folk Psychology as a Rough Template for Our Cognitive Profile: Some Problems

More generally, the perspective on cognition that emerges from neuroanatomy and neurophysiology holds out an entirely novel conception of the brain's fundamental mode of *representation*. The proposed new unit of representation is the *pattern of activation-levels* across a large population of neurons (*not* the internal sentence in some 'language of thought'). And the new perspective holds out a correlatively novel conception of the brain's fundamental mode of *computation* as well. Specifically, the new unit of computation is the *transformation of one activation-pattern into a second activation-pattern* by forcing it through the vast matrix of synaptic connections that one neuronal population projects to another population (*not* the manipulation of sentences according to 'syntactic rules'). Since our own dearly beloved folk psychology *shares* in classical AI's linguaformal portrayal of human cognitive activity, the new vector-coding/vector-processing portrayal of our cognitive processes therefore casts the integrity of folk psychology into doubt as well, at least as an account of the *basic* structure of cognitive activity. Point 4) of the preceding functionalist manifesto is therefore severely threatened, if not outright refuted, in addition to points 6) and 7). Its warm familiarity and yeoman social service notwithstanding, folk psychology appears to embody no insight whatever into the *basic* forms of representation and computation deployed by typical cognitive creatures.

This is an outcome that we should have expected in any case, since we appear to be the *only* species of cognitive creature on the planet that is capable of deploying the syntactic structures characteristic of language. If *all* cognition deploys them as the basic mode of doing business, why are the other terrestrial creatures so universally unable to learn any significant command of those linguistic structures? And if the basic modes of cognition in those other creatures are therefore almost certain to be *non*linguaformal in character, then why should we acquiesce in the delusion that human cognition – alone on the planet – *is* linguaformal in its basic character? After all, the human brain differs only marginally, in its microanatomy, from other mammalian brains; we are all closely proximate twigs on the same branch of the Earth's evolutionary tree. And the vector-coding/vector-processing story of how terrestrial brains do business is no less compelling for the human brain than it is for the brain of any other species. We have here a gathering case that folk psychology is a modern cousin of an old friend: Ptolemaic astronomy. It serves the purposes of rough prediction well enough, for an important but parochial range of phenomena. But it badly misrepresents what is really going on.[4]

IV. Multiple Realization: On the Alleged Impossibility of an Intertheoretic Reduction for Any Molar-Level Psychology

Conceivably, the preceding estimate of folk psychology is too harsh. Perhaps its presumptive failure to mesh with the vector-coding/vector-processing story of brain activity reflects only the fact that folk psychology is a molar-level portrait of cognitive activity, a portrait that picks up on laws and categories at a level of description far above the details of neuroanatomy and neurophysiology, a portrait that should not be *expected* to reduce to any such lower level of scientific theory. As many will argue, that reductive demand should not be imposed on folk psychology – nor on any potential replacement cognitive psychology either (a replacement drawn, perhaps, from future molar-level research). For, it will be said, psychology addresses lawlike regularities at its own level of description. These regularities are no doubt implemented in the underlying 'hardware' of the brain, but they need not be reducible to a theory of that hardware.[5] For there are endlessly many different possible material substrates that would sustain the same profile of molar-level cognitive activity.

The claim that molar-level cognitive activities are multiply realizable is almost certainly correct. Much less certain, however, is the idea that multiple realizability counts against the possibility of an intertheoretic reduction of folk psychology, and against the reduction of any scientific successor cognitive psychology that is similarly concerned with intelligence at the molar level. The knee-jerk presumption has always been that any such reduction to the underlying laws of any *one* of the many possible material substrates would be hopelessly *chauvinistic* in that it would automatically preclude the legitimate ascription of the cognitive

---

4 These skeptical themes go back a long way. See P. M. Churchland, "Eliminative Materialism and the Propositional Attitudes," *Journal of Philosophy* 78, no. 2 (1981): 67–90. For even earlier doubts, see P. K. Feyerabend, "Materialism and the Mind-Body Problem," *Review of Metaphysics* 17 (1963): 49–66; and R. Rorty, "Mind-Body Identity, Privacy, and Categories," *Review of Metaphysics* 19 (1965): 24–54.

5 Cf. J. A. Fodor, "The Special Sciences," 28 *Synthese* 28 (1974): 77–115.

vocabulary being reduced to entities composed of any of the many *other* possible material substrates. But this inference needs to be reexamined. It is, in fact, wildly fallacious.

What fuels the inference is the assumption that different material substrates – such as mammalian biology, invertebrate biology, extraterrestrial biology, semiconductor electronics, interferometric photonics, computational hydrology, and so on – will be governed by *different* families of physical laws. But this needn't be so. Let me illustrate with three salient and instructive examples.

Sound is a molar-level phenomenon. That is to say, it can be displayed only where there exists a large number of microscopic particles interacting in certain ways. And it, too, is a phenomenon that is multiply realized: in the Earth's highly peculiar atmosphere, in a gas of any molecular constitution, in a liquid of any molecular constitution, and in a solid of any molecular constitution. Sound propagates in any and all of these media. And yet sound is identical with, is smoothly reducible to, compression waves as propagated in any of these highly diverse media. For the underlying physical laws that bring the phenomenon of sound into the embrace of mechanical phenomena generally are *indifferent* to the peculiar molecules that make up the conducting medium, and to their collective status as a gas, liquid, or solid. What matters is that, collectively, those particles form an *elastic medium* that allows energy to be transmitted over long distances while the elements of the transmitting medium merely oscillate back and forth a comparatively tiny distance in the direction of energy transmission. To put it bluntly, the very *same* laws of wave propagation in an elastic medium cover *all* of the diverse cases at issue. Idiosyncratic features such as the *velocity* of wave propagation may indeed depend upon the details of the conducting medium (such as the mass of its molecules, and whether they form a gas, liquid, or solid). But the various high-level laws of acoustics (such as $v = \lambda\omega$, and other laws concerning the reflective and refractive behaviors of sound generally) reduce to the very same mechanical laws in all of these diverse cases. A diversity of material substrates here does *not* entail diversity in the underlying laws that govern those diverse substrates. Accordingly, acoustics is not an 'autonomous science', devoted to finding laws and ontological categories at its 'own level of description'. It is but one chapter in the broader mechanics of elastic media.

Temperature, also, is a molar-level phenomenon. And it, too, is a phenomenon that is multiply realized: in the Earth's atmosphere, or in any atmosphere, or indeed, in a gas of any molecular constitution whatever, either pure or mixed. For the temperature of a gas is identical with, is reducible to, the mean level of kinetic energy of the molecules that make up that gas. Here again, the underlying laws of motion (Newton's laws) that govern the behavior of, and the interactions of, the molecules involved are *the very same* for every kind of molecule that might be involved. Those laws are simply indifferent to the shape, or the mass, or the chemical makeup of whatever molecules happen to constitute the gas in question. Idiosyncratic details, such as the velocity of *dispersion* of an unconfined gas, will indeed depend on such details as molecular mass. But the laws of classical thermodymanics (such as the ideal gas law, $PV = \mu\mu RT$) reduce to the same set of underlying mechanical laws whatever the molecular makeup of the gas in question. Once again, a diversity of material substrates does *not* entail diversity in the underlying laws that govern those diverse substances. Accordingly, classical thermodynamics is not an 'autonomous science', devoted to finding laws and ontological categories at its 'own level of description'. Its reduction to statistical mechanics is a staple of undergraduate physics texts.

For a third example, a dipole magnetic field – as instanced in the simple rectangular bar magnet that one uses to pick up scattered thumb-tacks – constitutes a molar-level phenomenon, but such dipole magnetic fields are realizable in a variety of distinct metals and materials. Pure iron is the most familiar substrate, but sundry alloys (such as aluminum + nickel + cobalt) will also support such a field, as will certain metal/ceramic mixtures. Indeed, *any* substrate that somehow involves charged particles moving in mutually aligned circles (such as a tightly wound current-carrying coil of copper wire) will support a dipole magnetic field. For the simple laws that describe the shape and causal properties of such a field are all reducible to lower-level laws (Maxwell's equations) that describe the induction of electric and magnetic fields by the motion of charged particles such as electrons. And those lower-level laws are, once again, indifferent to the details of whatever material substrate happens to sustain the circular motion of charged particles.

Once again, an open-ended diversity of sustaining substrates does not entail the irreducibility of the molar-level phenomenon therein sustained. And the historical pursuit of the various pre-Maxwellian theories of dipole magnetic fields (e.g., 'effluvium' theories) did not constitute an 'autonomous science', forever safe from the reductive reach of new and more comprehensive theories. On the contrary, the work of Faraday and Maxwell brought those older theories into the welcoming embrace of the new, and much to the illumination of the former.

These examples can be multiplied. But even one such example illustrates the hazards of inferring the irreducibility of some molar-level phenomenon from the premise of its multiple realizability, even when that premise happens to be *true*. For diverse material substrates may still be governed, all of them, by some of the very *same* low-level physical laws, laws quite capable of explaining the molar-level behaviors *shared* by all of those diverse substrates. The classical argument for point 5) of the functionalist manifesto is plainly flawed, for it naïvely ignores these obvious sorts of reductive possibilities.

Let us push this line of thought a little further. For it now begins to appear that, beyond merely providing decisive counterexamples to what functionalism presents as a robustly *non*reductive pattern, the three examples just cited also provide an *alternative* pattern – a pattern whereby molar-level theories that record genuine high-level regularities across diverse material substrates will *typically* find a successful reduction to some underlying and highly general physical laws, laws that are simply blind to the idiosyncratic and irrelevant differences that happen to distinguish the several substrates. We have just seen this happen in three unproblematic cases. And there are, I repeat, many more.[6]

---

[6] For a fourth example, consider the shared molar-level thermodynamic profile of living organisms across a wide variety of biochemical substrates. The underlying laws of nonequilibrium thermodynamics are once again blind to the peculiar chemical makeup of such diverse substrates. For a fifth example, consider Kepler's three laws of planetary motion, valid for planets of highly diverse material constitution. All three of those laws are reducible to Newton's particle mechanics plus his universal law of gravitation. For a sixth example, consider the science of aerodynamics: the theory of creatures or machines that are capable of flight. Multiple realizability is an obvious feature of this domain: think of seagulls, hummingbirds, bats, dragonflies, wooden airplanes, metal airplanes, helicopters, and so on. And yet their shared molar behavior is ultimately owed to the fact that they all contrive to accelerate ambient air more-or-less continuously downward, which activity yields, by Newton's (substrate-neutral) third law, a reactive upward force that is more-or-less continuously equal to the task of keeping them aloft. For a seventh example, consider the closely similar chemical and electrical behaviors of the distinct elements within a given chemical 'family', those that constitute one vertical column of the periodic table (e.g., the metals, or, the noble gases). Here the shared molar-level chemical regularities, across a given family of elements, are explained in terms of shared valence–electron shell structures across the distinct types of atoms within that family. For an eighth example, one rather closer to the case of cognitive creatures, consider the molar behavior of any music player, radio, or TV. Despite the great variety of metal and semiconductor substrates that will instantiate the required circuits for signal detection, amplification, and presentation, the behavior of all such devices is reducible to the same set of electrodynamical laws concerning resistances, capacitances, and inductances – laws blind to the material diversity of the substrates that a given manufacturer may choose to employ. As we see from such examples, this general *reductive* pattern, across substrate

On this alternative logical and historical pattern, legitimate molar-level theories that comprehend genuine natural kinds will thus be positively *expected* to find some such intertheoretic reduction. For if they eventually prove *not* to be thus reducible, we will have to reconsider the initial presumption that the molar-level theory really does embrace genuine high-level natural kinds governed by genuine high-level explanatory laws. The 'unitary' account that the molar theory seemed to provide, across the diverse substrates, might then have to be judged an accidental or a false unity. And its supposedly lawlike generalizations will thus turn out to be *accidental* generalizations of some sort, generalizations that are empty of real explanatory and predictive power. Accordingly, if we expect our beloved folk psychology, or *any* psychology, to provide an accurate, natural-kind–embracing, genuinely nomological and explanatory account of the molar-level cognitive operations and behavior of humans, other mammals, humanlike aliens, and humanlike artificial automata, then we had better *hope* that there exist highly general underlying laws – laws blind to the material differences among all of these diverse creatures – that serve collectively to explain, and thus to reduce, the categories and laws of psychology.

Let us finally confront the most important question here at issue. Just what are the chances of finding some substrate-neutral underlying laws – laws with a suitably broad explanatory reach – for *psychology* in particular? That is to say, what are the chances that the case of psychology will turn out to be an instance of the alternative and overtly *reductive* pattern of development explored in the preceding pages, and in the examples of footnote 6? Well, they are certainly not zero. For there are at least two low-level theories that have sufficient generality to embrace all of the diverse material realizations of cognitive activity listed in the preceding paragraph, and that also hold promise for explaining at least some of the activities comprehended by psychology. Let us take a look at them.

V. Some Reductive Possibilities for Molar-Level Psychology

The first possible framework equal to the task of comprehending all of the diverse material realizations envisioned for cognitive systems is nonequilibrium thermodynamics. Distinct from the more familiar

---

diversity, is quite robust. For an illuminating discussion, see M. Strevens, *Bigger than Chaos* (Cambridge, MA: Harvard University Press, 2003), especially chap. 5, "Implications for the Philosophy of the Higher-Level Sciences."

(near-equilibrium) statistical thermodynamics discussed earlier, this is the general framework for describing the laws of energy and information flow in partially closed physical systems that are, and remain, very far away from energetic equilibrium. This is the framework, still in its developmental infancy, that already unites and illuminates all *biological* phenomena, whatever their physical constitution. The basic idea, first outlined half a century ago by the physicist Erwin Schrodinger,[7] is that any living organism is a highly improbable physical structure whose natural behavior – if it is located in a suitable flow of ambient energy – serves to exploit whatever structure it already contains so as to produce *additional* physical structure. It grows, or it repairs itself, or it reproduces. Such an interest-bearing investment[8] is possible only when the system is situated so as to exploit an energy flux that begins with energy from a very low-entropy state,[9] energy that is then progressively dissipated into energy at a much higher entropy state. The living physical system 'steals' some of the initial low-entropy energy as that energy courses through it, and it then incorporates that energy in the form of additional (and improbable) physical structure. The low-entropy energy source for our terrestrial environment is ultimately the Sun, radiating at a black-body temperature of roughly 4000 K (i.e., at rather short wavelengths). And the ultimate high-entropy energy sink is the surrounding background of empty space, radiating at a black-body temperature of about 3 K (i.e., at very long wavelengths). In between lies the biosphere at a temperature around 293 K. Without such a concentrated or low-entropy energy source 'above' us, and such a dissipated high-entropy energy sink 'below' us, nothing alive could hope to remain alive. Indeed, without such conditions nothing of any biological interest – that is, no extremely improbable physical structures with complex metabolic pathways – could ever have evolved in the first place.

Those that have evolved are thus instant testaments to the existence of such a complexity-inducing ambient energetic waterfall – a constant flow from the Sun, through us, and into the cold abyss beyond. Moreover, any individual of any species also embodies, in its typical structural details, extraordinary amounts of information about the peculiar environmental niche in which it thrives. For no individual could be expected

---

[7] E. Schrodinger, "What Is Life?" (Cambridge: Cambridge University Press, 1944).

[8] Note well the economic metaphor here deployed. Its aptness will come up again shortly.

[9] Entropy is a measure of how chaotically *dissipated* an amount of energy happens to be, a measure of how *unavailable* it is to do work, a measure (if you like) of its *weariness*. By contrast, a low-entropy state implies energy that is highly 'concentrated' and available to do work.

to *have* the specific physical structure it has *unless* the environment in which it thrives has a comparably specific physical and dynamical profile. For, once again, eons of evolutionary pressures have made the former exquisitely 'tuned' to the latter, functionally and metabolically speaking. If you want an indirect but highly informative window onto the chemical and biological dynamics of a forest pond, examine the frog who lives there.

I provide the reader with this brief sketch, of how nonequilibrium thermodynamics embraces biological phenomena generally, not just to validate the claim made in footnote 6, but to prepare the ground for the following suggestion. Cognitive phenomena are just an additional instance or iteration of the thermodynamical profile already outlined. Specifically, a creature that *learns* about the world is a creature that exploits the low-entropy internal structure or information that it already possesses, in such a fashion that, if the creature is placed in an environment with an information-rich energy flow, it comes to embody *additional* information-bearing structure about its environment, typically in the form of a progressively rewired brain.

The relevant energy flow here begins with the low-entropy states of incoming *sensory* signals (light, sound, pressure, taste, smell, whatever), signals that contain detailed information about the immediate physical environment. And it ends with the dissipation of that original low-entropy energy, after the brain's cognitive processing is done, as high-entropy heat radiated away by the body at long wavelengths in the infrared, its original information 'lost', or rather, left behind in the brain. The active cognitive system 'steals' some of the low-entropy energy that its sensory organs provide, and incorporates it as additional information-bearing structure. My biological body at age six days will embody a great detail of general information about my natural environment, as we noted two paragraphs ago, concerning the frog. But my brain, at six years (or at six decades), embodies an additional wealth of information, information that the human genome was, and is, far too poor to bequeath to me. I have to acquire that wealth of information postnatally. And I do. And so does a mouse. And so does any cognitive creature. For that is what a cognitive creature is: an 'extrasomatic information multiplier'. Unlike a typical heat engine (e.g., a steam engine, an automobile engine), which exploits an entropy-increasing energy flow to produce macroscopic *motion*, the brain exploits such a flow to produce neuronally and synaptically embodied *information*. We are, in fact, *epistemic* engines, not just figuratively, but literally.

This naturalistic portrait, note well, makes no reference to the variety of material substrates that might sustain such an energetic and informational economy. Many different substrates are presumably possible. What makes them all cognitive creatures – part of it, anyway – is their shared thermodynamic and information-multiplying profile.

I briefly floated this possible construal of cognitive creatures in 1979, in the closing paragraphs of my book *Scientific Realism and the Plasticity of Mind*,[10] and again, in slightly more detail, in a 1982 paper.[11] Those accounts of cognition, and that of the preceding two paragraphs, however, may well be dismissed as mere hand-waving speculation, unless we can provide an account of *how* brains actually process, and incorporate into their internal structure, ambient information.

In the salient case of *biological* metabolisms, we do indeed possess such a non-hand-waving account. We know how DNA embodies information. We know how that information is read out by RNA in order to synthesize various protein molecules. We know how those protein molecules catalyze certain metabolic reactions and sequences of such reactions. We know how those reaction chains create new biological molecules that form additional biological structures. We know how those structures collectively steer ambient energy and materials along paths that sustain and amplify the organism at issue. The nonequilibrium thermodynamical portrait of living things is therefore not just a philosophical guess. It is a highly general reductive framework that brings real illumination to biological processes, across a wide diversity of chemical substrates.

Twenty-five years ago, I must own, the nonequilibrium thermodynamical portrait of *cognitive* activity *was* a merely philosophical guess – a hesitant extrapolation from the thermodynamic portrait of living things just explored. For we then lacked any corresponding account of how brains actually process and eventually store new information. In the intervening years, however, an account of exactly those activities has pieced itself together and has become the focus of a great deal of research, both experimental and theoretical. That account posits fleeting *activation vectors* across proprietary populations of neurons as the basic mode of ephemeral representation. It posits *vector-to-vector transformations*, at the hands of intervening matrices of synaptic connections, as the basic mode of computation over those representations. It posits specific *configurations of weighted synaptic connections* as

---

[10] (Cambridge: Cambridge University Press, 1979).
[11] "Is *Thinker* a Natural Kind?" *Dialogue* 21, no. 2 (1982): 223–38.

the basic mode of general or background knowledge. And it posits ongoing *adjustments in the values* of those weighted synaptic connections as the most basic mode of learning.[12]

It is vital to appreciate that the structural and dynamical portrait just painted – of vector coding and vector processing via large matrices with plastic coefficients – is once again a portrait that can be realized in a wide variety of material substrates: in mammalian brains, in octopus brains, in extraterrestrial brains, in electronic chips, in optical systems, and so forth. For the mathematical laws of vector/matrix processing, and of the information-dependent, experience-driven adjustments of the individual coefficients of the transforming matrix, are all indifferent to the physical medium that displays those fleeting vectors and embodies those comparatively enduring matrices. Some idiosyncratic details, such as cognitive reaction times, will indeed be sensitive to the implementational facts, such as the speed of vectorial conduction between distinct populations of active units. (As noted earlier, an electric current in a copper wire propagates much faster than a spike train in an axon.) But the profile of molar-level activity will be importantly similar across all of these diverse substrates.

Once again we are contemplating a low-level explanation, in terms of general or abstract underlying natural laws, of a roughly constant profile of molar-level activity, activity that can be displayed across a considerable variety of material substrates. But this time the explanatory target is the *cognitive* activities of creatures like ourselves. We are no longer pressing a mere a priori possibility on our functionalist friends. We are confronting a pair of real theories (the vector/matrix account of brain structure and function, and the nonequilibrium thermodynamical account of brain growth and learning), theories that hold some nontrivial promise of providing systematic reductive explanations of cognitive phenomena in particular, despite their presumably diverse realizations. To put it bluntly, we are confronting exactly what classical functionalism said was not to be had, nor even to be sought.

These new developments, especially the vector/matrix story, have already given us a much deeper understanding of what the brain does and of how it manages to do it. We now understand, for example, how the

---

[12] For accessible, entry-level accounts that will provide doors to the wider literature, see P. S. Churchland and T. J. Sejnowski, *The Computational Brain* (Cambridge, MA: MIT Press, 1987); P. M. Churchland, *The Engine of Reason, the Seat of the Soul: A Philosophical Journey into the Brain* (Cambridge, MA: MIT Press, 1995); Churchland, "Inner Spaces and Outer Spaces."

activation space of a large population of neurons can come to embody a structured system of categorical prototypes – that is, a meaningful *conceptual framework*. We now understand how those prototype-points in such a background neuronal activation space can be selectively activated by sensory inputs. That is, we understand how a brain can *interpret* its sensory experience in terms of its acquired conceptual framework. We now understand how prototypical motor behaviors can be represented as prototypical activation-trajectories in motor-neuron activation space. That is, we have some understanding of how complex *motor skills* are embodied. And we know how such unfolding trajectories can actually *generate* the relevant motor behaviors in the body's limb and muscle systems. In sum, we can now see cognitive activity as we have never seen it before. Whether we are seeing it correctly, only time will tell. But a fertile vision is already being explored. What this means is that the celebrated point 5) of the functionalist manifesto is not just naïvely argued. In fact, it is almost certainly false.[13]

VI. What Does *Not* Unite the Class of Cognitive Creatures

On the vector/matrix story explored in the preceding section, what carries the burden of any creature's acquired background knowledge, of the world's general and enduring structure, is the specific configuration of the billions or trillions of synaptic connections that variously intervene between the brain's many distinct neuronal-coding populations. It is these variously weighted excitatory or inhibitory elements that constitute the brain's principal memory store, and also its principal means of computation. One and the same system simultaneously serves both functions.

---

[13] Allow me a closing remark on Jerry Fodor's 1974 parade case of a molar-level natural science for which reductive aspirations are supposed to be clearly foolish, namely, economics. The supporting argument then appealed to the multiple realizability of *currency* systems – shell currency, coin currency, paper currency, electronic currency, and so on. We can all agree that economics is not going to be reducible to the chemistry of wood fiber, or to the physics of copper and gold. But all of this is now visibly beside the point. For we can now appreciate that economics is the study of *the metabolisms of superorganisms*, a phenomenon that once again falls firmly within the province of nonequilibrium thermodynamics, a science whose laws are blind to such implementational details concerning currency. A national economy, after all, embodies a flow of both energy and materials: it creates real physical and organizational structures, and it dissipates vast amounts of (initially low-entropy) energy in the process. It is too soon to *insist* that economics will indeed find such an explanatory reduction. But neither can Fodor justly insist that it won't. The presumption in favor of his principal nonreductive example has just evaporated.

No two people, however, display the same configuration of synaptic connections and synaptic weights. Each human brain boasts roughly 10^{14} synaptic connections, the overwhelming majority of which are established postnatally in response to a lived experience that is unique to each individual. Since we experience a common world that does display enduring features, each of us ends up with a family of sculpted activation-spaces whose structure is *similar* to the structure of other people's activation spaces, at least if they are members of the same culture. But genuine identity is too much to ask for. We may all agree that the vector/matrix system found in each individual is computing a function of some fabulous complexity. But no two people on the planet will be computing exactly the same function, for no two people share the same matrix of synaptic connections.

Very well, but surely they will be computing *similar* functions? Indeed so, if they happen to be peas from the same cultural pod.[14] But what wants emphasizing here is the real functional *diversity* displayed by individuals at different stages of their lives, in different cultures across the planet, and at different points in our very long and cognitively diverse human history. This diversity in the functions that brains are computing becomes more striking still when we expand our consideration to include cognitive activity in the nervous systems of other terrestrial creatures such as chimpanzees, cats, mice, finches, crocodiles, crabs, octopi, fish, and spiders. Clearly such diverse creatures are not all computing the same function, nor even remotely similar functions. And yet, we are all cognitive creatures.

What is it, then, that unites us? Ironically, it appears to be the abstract form of our *hardware* that unites us! We – all of us on the preceding list – are massively parallel vector-processors whose ever-active vector-transforming matrices (our trillions of synaptic connections) are slowly updated or instructed by a procedure that filters information from a low-entropy flux of energy from our sensory peripheries. This computational arrangement has prodigious advantages over the serial architecture deployed in classical (von Neumann) computers – in its speed of computation, in its graceful tolerance of scattered component failures, and

---

[14] How this similarity can be achieved, despite our synaptic diversity, is detailed in P. M. Churchland, "Conceptual Similarity across Sensory and Neural Diversity: The Fodor/Lepore Challenge Answered," *Journal of Philosophy* 95, no. 1 (Jan., 1998): 5–32. See also Chapter 8 of this volume.

in its swift deployment of relevant information. This alternative computational template is sufficiently virtuous to make it a likely evolutionary attractor on any planet that develops life, not just on Earth, and to make it a compelling technological choice for any future attempts at constructing artificial intelligence as well.

Accordingly, point (1) of the original functionalist manifesto is almost certainly a mistake – indeed, a monumental mistake. The cognitive creatures on this planet are computing a bewildering *variety* of very different functions, but they are all using fundamentally similar computational 'hardwares' to do it. On this fundamental point, classical functionalism had things exactly backward.

Point (2) must therefore be rejected as well. If our alternative portrait of cognition is even roughly correct, the central job of cognitive psychology is to explore *how* it is that terrestrial brains are able to compute the extraordinary *variety* of functions displayed in diverse species of cognitive creatures. This must be an empirical undertaking, one sensitive to the idiosyncrasies of nonhuman nervous systems. Accordingly, point (3) must be rethought along the same lines. The central job of AI research is not just to explore the construction of artificial vector-processing systems that compute the same function that some species of animal is already computing. A central part of its job will be to explore instead the pregnant potential of such artificial systems for computing functions – for pursuing cognitive activities – that no terrestrial creature has yet pursued or ever will pursue. Large-scale electronic realizations of our vector/matrix style of computational resources will explore entirely new horizons for information processing and world representation, and, being electronic, they will do it roughly a million times faster than biological creatures can ever hope to do it (because the speed of signal conduction in a copper wire is close to the speed of light). The enterprise of artificial intelligence thus has a dazzling future, but not because classical functionalism launched it in the right direction.

Indeed, it launched the enterprise in a most unfortunate direction. Point (6) provided a twisted rationale for mostly *ignoring* the empirical or experimental neurosciences, and for ignoring the early theoretical work that attempted to model the activities of large numbers of interconnected neurons. Worse still, point (7) celebrated this deliberate disconnection with an ill-conceived positive portrait of cognitive psychology and artificial intelligence as 'methodologically autonomous sciences'. In retrospect, this was unwise, despite the genuinely clever contributions of a great many gifted researchers. For it served to insulate the relevant research from exactly the empirical information that promised the most interesting and authoritative constraints on whatever models were put up for evaluation. The result was almost half a century of misdirected research.

VII. Conclusions

The seven elements of our opening functionalist manifesto, all seven of them, appear to be false – not just inadequately argued for, but outright false. Fortunately, we now have an alternative positive portrait of cognitive activity in place, one capable of steering systematic research on the nature of cognition, research that points in directions interestingly different from the directions that dominated the last half of the twentieth century. That alternative portrait is no longer imprisoned by the linguaformal representational paradigm provided by folk psychology, nor is it centered on the computational paradigm provided by the digital-coding, serial-processing, program-running machines of the still-standard von Neumann configuration. Instead, it draws its opening inspirations, in both cases, from the deeply instructive empirical example of terrestrial nervous systems.

In closing, it is worth pointing out that two prominent background assumptions of the functionalist program have not been denied in the preceding critique. The first assumption is that cognitive creatures are indeed engaged in computing complex functions of some sort or other. And the second is that these computational activities, whatever they are, can be realized in a diversity of physical substrates. These assumptions are presumably as true, and as important, as they ever were. But in the present intellectual environment, those same two assumptions now pull our imaginations in entirely new directions. The first assumption motivates the brain-centered research program known as computational neurobiology. And the second assumption motivates the development of alternative physical realizations (presumably electronic or photonic) – not of our 'software' (strictly speaking, we don't *have* any!), but – of the massively parallel, vector-processing structure of our biological *hardware*. Let us hope that this second wave of research will be more revealing, and less self-blinkered, than the functionalist-inspired wave that preceded it.

3

Toward a Cognitive Neurobiology of the Moral Virtues

I. Introduction

These are the early days of what I hope will be a long and fruitful intellectual tradition, a tradition fueled by the systematic interaction and mutual information of cognitive neurobiology on the one hand and moral theory on the other. More specifically, it is the traditional subarea we call metaethics, including moral epistemology and moral psychology, that will be most dramatically informed by the unfolding developments in cognitive neurobiology. And it is metaethics again that will exert a reciprocal influence on future neurobiological research – more specifically, into the nature of moral perception, the nature of practical and social reasoning, and the development and occasional corruption of moral character.

This last point about reciprocity highlights a further point. What we are contemplating here is no imperialistic takeover of the moral by the neural. Rather, we should anticipate a mutual flowering of both our high-level conceptions in the domain of moral knowledge and our lower-level conceptions in the domain of normal and pathological neurology. For each level has much to teach the other, as this essay will try to show.

Nor need we resist this interaction of distinct traditions on grounds that it threatens to deduce normative conclusions from purely factual premises, for it threatens no such thing. To see this clearly, consider the following parallel. Cognitive neurobiology is also in the process of throwing major illumination on the philosophy of *science* – by way of revealing the several forms of neural representation that underlie scientific cognition, and the several forms of neural activity that underlie learning and conceptual change (see, e.g., Churchland 1989a, chaps. 9–11). And yet,

substantive science itself will still have to be done by scientists according
to the various methods by which we make scientific progress. An adequate
theory of the brain, plainly, would not constitute a theory of stellar evolution or a theory of the structure of the periodic table. It would constitute,
at most, only a theory of how we generate, embody, and manipulate such
worthy cognitive achievements.

Equally, and for the same reasons, substantive moral and political theory will still have to be done by moral and political thinkers, according
to the various methods by which we make moral and political progress.
An adequate theory of the brain, plainly, will not constitute a theory of
distributive justice or a body of criminal law. It would constitute, at most,
only a theory of how we generate, embody, and manipulate such worthy
cognitive achievements.

These reassurances might seem to rob the contemplated program
of its interest, at least to moral philosophers, but we shall quickly see
that this is not the case. For we are about to contemplate a systematic
and unified account, sketched in neural-network terms, of the following
phenomena: moral knowledge, moral learning, moral perception, moral
ambiguity, moral conflict, moral argument, moral virtue, moral character, moral pathology, moral correction, moral diversity, moral progress,
moral realism, and moral unification. This collective sketch will serve
at least to outline the program, and even at this early stage it will provide a platform from which to address the credentials of one prominent
strand in preneural metaethics, the program of so-called virtue ethics, as
embodied in both an ancient writer (Aristotle) and three modern writers
(Johnson, Flanagan, and MacIntyre).

II. The Reconstruction of Moral Cognitive Phenomena
in Cognitive Neurobiological Terms

This essay builds on work now a decade or so in place, work concerning
the capacity of recent neural-network models (of microlevel brain activity) to reconstruct, in an explanatory way, the salient features of molar-
level cognitive activity. That research began by addressing the problems
of perceptual recognition, motor-behavior generation, and other basic
phenomena involving the gradual learning of sundry cognitive *skills* by
artificial "neural" networks, as modeled within large digital computers
(Gorman and Sejnowski 1988b; Lehky and Sejnowski 1988; Rosenberg
and Sejnowski 1987; Lockery, Fang, and Sejnowski 1991; Cottrell 1991;
Elman 1992). From there, it has moved both downward in its focus, to

try to address in more faithful detail the empirical structure of biological brains (Churchland and Sejnowski 1992), and upward in its focus, to address the structure and dynamics of such higher-level cognitive phenomena as are displayed, for example, in the human pursuit of the various theoretical sciences (Churchland 1989a).

For philosophers, perhaps the quickest and easiest introduction to these general ideas is the highly pictorial account in Churchland (1995b), to which I direct the unprepared reader. My aim here is not to recapitulate that groundwork, but to build on it. Even so, that background account will no doubt slowly emerge here, from the many examples to follow, even for the reader new to these ideas, so I shall simply proceed and hope for the best.

The model here being followed is my earlier attempt to reconstruct the epistemology of the *natural* sciences in neural-network terms (Churchland 1989a). My own philosophical interests have always been centered on issues in epistemology and the philosophy of science, and so it was natural, in the mid-1980s, that I should first apply the emerging framework of cognitive neurobiology to the issues with which I was most familiar. But it soon became obvious to me that the emerging framework had an unexpected generality, and that its explanatory power, if genuine, would illuminate a much broader range of cognitive phenomena than had so far been addressed. I therefore proposed to extend its application into other cognitive areas such as mathematical knowledge, musical knowledge, and moral knowledge. (Some first forays appear in chapters 6 and 10 of Churchland 1995b.) These further domains of cognitive activity provide, if nothing else, a series of stiff *tests* for the assumptions and explanatory ambitions of neural-network theory. Accordingly, the present paper presumes to draw out the central theoretical claims, within the domain of metaethics, to which a neural-network model of cognition commits us. It is for the readers, and especially for those who are professional moral philosophers, to judge for themselves whether the overall portrait that results is both explanatorily instructive and faithful to moral reality.

1. Moral Knowledge

Broadly speaking, to teach or train any neural network to embody a specific cognitive capacity is gradually to impose a specific *function* onto its input–output behavior. The network thus acquires the ability to respond, in various but systematic ways, to a wide variety of potential sensory inputs. In a simple, three-layer *feedforward* network with fixed synaptic connections (Figure 3.1*a*), the output behavior at the third layer of neurons

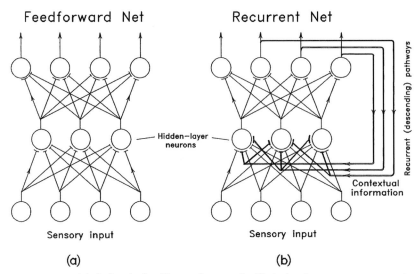

FIGURE 3.1. (*a*) A simple feedforward network. (*b*) A simple recurrent network. For a quick grip on the functional significance of such models, think of the lower or input layer of neurons as the sensory neurons, and think of the upper or output layer of neurons as the motor or muscle-driving neurons.

is completely determined by the activity at the sensory input layer. In a (biologically more realistic) *recurrent* network (Figure 3.1*b*), the output behavior is jointly determined by sensory input *and* by the prior dynamical state of the entire network. The former case yields a cognitive capacity that is blind to temporal context; the latter yields a capacity that is sensitive to, and responsive to, the changing cognitive contexts in which its sensory inputs are variously received. In both cases, the acquired cognitive capacity actually resides in the specific configuration of the many synaptic connections between the neuronal layers, and *learning* that cognitive capacity is a matter of slowly adjusting the size or "weight" of each connection so that, collectively, they come to embody the input–output function desired. On this, more in a moment.

Evidently, a trained network has acquired a specific skill. That is, it has learned how to respond, with appropriate patterns of neural activity across its output layer, to various inputs at its sensory layer. Accordingly, and as with all other kinds of knowledge, my first characterization of moral knowledge portrays it as a *set of skills*. To begin with, a morally knowledgeable adult has clearly acquired a sophisticated family of *perceptual* or *recognitional* skills, which skills allow him a running comprehension of his own social and moral circumstances, and of the social and moral circumstances of the others in his community. Equally clearly, a morally knowledgeable adult has acquired a complex set of *behavioral* and *manipulational* skills, which skills make possible his successful social and moral interaction with the others in his community.

According to the model of cognition here being explored, the skills at issue are embodied in a vast configuration of appropriately weighted synaptic connections. To be sure, it is not intuitively obvious how a thousand, or a billion, or a trillion such connections can constitute a specific cognitive skill, but we begin to get an intuitive grasp of how they can do so when we turn our attention to the collective behavior of the neurons at the layer to which those connections happen to attach.

Consider, for example, the second layer of the network in Figure 3.1*a*. That neuronal population, like any other discrete neuronal population, represents the various states of the world with a corresponding variety of *activation patterns* across that entire population. That is to say, just as a pattern of brightness levels across the 200,000 pixels of your familiar TV screen can represent a certain two-dimensional scene, so can the pattern of activation levels across a neuronal population represent specific aspects of the external world, although the "semantics" of that representational relation will only rarely be so obviously "pictorial." If the neuronal representation is auditory, for example, or olfactory, or gustatory, then obviously the representation will be something other than a two-dimensional "picture."

What is important for our purposes is that the abstract *space* of *possible* representational patterns, across a given neuronal population, slowly acquires, in the course of training the synapses, a specific structure – a structure that assigns a family of dramatically preferential abstract *locations*, within that space, in response to a preferred family of distinct stimuli at the network's sensory layer. This is how the mature network manages to categorize all possible sensory inputs: either as instances of one or the other of its learned family of *prototypical categories*, or, failing that, as instances of unintelligible noise. Before training, *all* inputs produce noise at the second layer. After training, however, that second layer has become preferentially sensitized to a comparatively tiny subset of the vast range of possible input patterns (most of which are never encountered). Those "hot-button" input patterns, whenever they occur, are subsequently assimilated to the second layer's acquired set of *prototypical categories*.

Consider an artificial network (Figure 3.2*a*) trained to discriminate human faces from nonfaces, male faces from female faces, and a handful of named individuals as presented in a variety of distinct photographs. As a result of that training, the abstract space of *possible* activation patterns

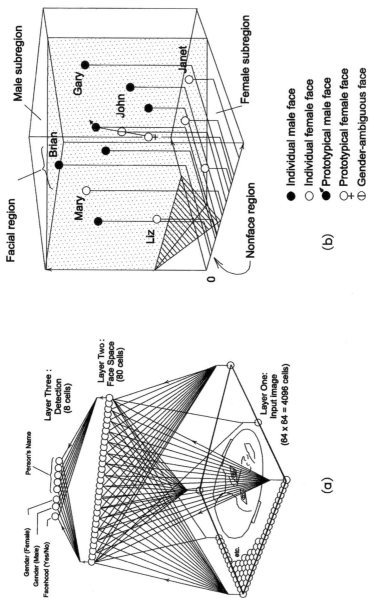

FIGURE 3.2. (a) A feedforward neural network for recognizing human faces and distinguishing gender. (b) The hierarchy of categorial partitions, acquired during training, across the space of possible neuronal activation patterns at the network's middle or "hidden" layer. This three-dimensional cube is a cartoon; the real space has eighty dimensions, one for each neuron at the middle layer.

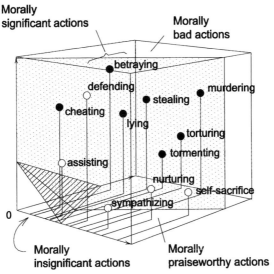

FIGURE 3.3. A (conjectural) activation space for moral discrimination.

across its second neuronal layer has become *partitioned* (Figure 3.2*b*), first into a pair of complementary subvolumes for neuronal activation patterns that represent sundry faces and nonfaces, respectively. The former subvolume has become further partitioned into two mutually exclusive subvolumes for male faces, and female faces, respectively. And within each of these two subvolumes, there are proprietary "hot spots" for each of the named individuals that the network learned to recognize during training.

Following this simple model, the suggestion here advanced is that our capacity for *moral* discrimination also resides in an intricately configured matrix of synaptic connections, which connections also partition an abstract conceptual space, at some proprietary neuronal layer of the human brain, into a hierarchical set of categories, categories such as "morally significant" versus "morally nonsignificant" actions; and within the former category, "morally bad" versus "morally praiseworthy" actions; and within the former *sub*category, sundry specific categories such as "lying," "cheating," "betraying," "stealing," "tormenting," "murdering," and so forth (Figure 3.3). That abstract space of possible neuronal-activation patterns is our conceptual space for moral representation, and it displays an intricate structure of similarity and dissimilarity relations, relations that cluster similar vices close together and similar virtues close together, relations that separate highly dissimilar action categories into spatially distant sectors of the space. This high-dimensional similarity space displays a structured family of categorical "hot spots" or "prototype positions," to which actual sensory inputs are assimilated with varying degrees of closeness.

An abstract space of *motor-neuron* activation patterns will serve a parallel function for the generation of actual social behavior, a neuronal layer that presumably enjoys close functional connections with the sensory neurons just described. All told, these structured spaces constitute our acquired knowledge of *the structure of social space,* and *how to navigate it effectively.*

2. *Moral Learning*

Moral learning consists in the gradual generation of these internal perceptual and behavioral prototypes, a process that requires repeated exposure to, or practice in, various *examples* of the perceptual or motor categories at issue. In artificial neural networks, such learning consists in the repeated adjustment of the weights of their myriad synaptic connections, adjustments that are guided by the naïve network's initial performance *failures,* as measured by a distinct "teacher" program. In biological creatures, too, learning consists in the repeated adjustment of one's myriad synaptic connections, a process that is also driven by one's ongoing experience with failure. Our artificial "learning technologies" are currently a poor and pale reflection of what goes on in real brains, but in both cases – the artificial networks and the real brains – those gradual synaptic readjustments lead to an appropriately structured high-dimensional similarity space, a space partitioned into a hierarchical family of categorical subspaces, which subspaces contain a central hot spot that represents a *prototypical* instance of its proprietary category.

Such learning typically takes time, often large amounts of time. And as the network models have also illustrated, such learning often needs to be structured, in the sense that the simplest of the relevant perceptual and behavioral skills need to be learned first, with the more sophisticated skills being learned later, and only after the elementary ones are in place. Moreover, such learning can display some familiar pathologies, those that derive from a narrow or otherwise skewed population of training examples. In such cases, the categorical framework duly acquired by the network fails to represent the full range and true structure of the domain it needs to represent, and performance failures are the inevitable result.

These remarks barely introduce the topic of moral learning, but we need to move on. The topic will be readdressed later, when we discuss moral progress.

3. Moral Perception

This most fundamental of our moral skills consists in the *activation*, at some appropriate layer of neurons at least half a dozen synaptic connections away from the sensory periphery, of a specific *pattern* of neuronal excitation-levels that is sufficiently close to some already learned moral pattern. That *n*th-layer activation pattern is jointly caused by the current activation pattern across one or more of the brain's sensory or input layers, and by the series of carefully trained synaptic connections that intervene. Moral perception is thus of a piece with perception generally, and its profile displays features long familiar to perceptual psychologists.

For example, our spontaneous judgments about the social and moral configuration of our current environment are strongly sensitive to contextual features, to collateral information, and to our current interests and focus of attention. Moral perception is thus subject to "priming effects" and "masking effects," just as in perception generally. As well, moral perception displays the familiar tendency of cognitive creatures to "jump to conclusions" in their perceptual interpretations of partial or degraded perceptual inputs. Like artificial networks, we humans have a strong tendency to automatically assimilate our current perceptual circumstances to the *nearest* of the available moral prototypes that our prior training has created in us.

4. Moral Ambiguity

A situation is morally ambiguous when it is problematic by reason of its tendency to activate *more than one* moral prototype, prototypes that invite two incompatible or mutually exclusive subsequent courses of action. In fact, and to some degree, ambiguity is a chronic feature of our moral experience, partly because the social world is indefinitely complex and various, and partly because the interests and collateral information each of us brings to the business of interpreting the social world differ from person to person and from occasion to occasion. The recurrent or descending pathways within the brain (illustrated, in stick-figure form, in Figure 3.1*b*) provide a continuing stream of such background information (or misinformation) to the ongoing process of perceptual interpretation and prototype activation. Different "perceptual takes," on one and the same situation, are thus inevitable. Which leads us to our next topic.

5. Moral Conflict

The activation of distinct moral prototypes can happen in two or more distinct individuals confronting the same situation, and even in a single individual, as when some contextual feature is alternatively magnified or minimized and one's overall perceptual take flips back and forth between two distinct activation patterns in the neighborhood of two distinct prototypes. In such a case, the single individual is morally conflicted ("Shall I *protect* a friend's feelings by keeping silent on someone's trivial but hurtful slur, or shall I be forthright and *truthful* in my disclosures to a friend?").

Interpersonal conflicts about the moral status of some circumstance reflect the same sorts of divergent interpretations, driven this time by interpersonal divergences in the respective collateral information, attentional focus, hopes and fears, and other contextual elements that each perceiver brings to the ambiguous situation. Occasional moral conflicts are thus possible – indeed, they are inevitable – even between individuals who had identical moral training and who share identical moral categories.

There is, finally, the extreme case where moral judgment diverges because the two conflicting individuals have fundamentally different moral conceptual frameworks, reflecting major differences in the acquired structure of their respective activation spaces. Here, even communication becomes difficult, and so does the process by which moral conflicts are typically resolved.

6. *Moral Argument*

On the picture here being explored, the standard conception of moral argument as the formal deduction of moral conclusions from shared moral premises starts to look procrustean in the extreme. Instead, the administration and resolution of moral conflicts emerges as a much more dialectical process whereby the individuals in conflict take turns highlighting or making salient certain aspects of the situation at issue, and take turns urging various similarities between the situation at issue and various shared prototypes, in hopes of producing, within their dialectical adversary, an activation pattern that is closer to the prototype being defended ("It's a mindless clutch of cells, for heaven's sake! The woman is not obliged to preserve or defend it.") and/or further from the prototype being attacked ("No, it's a miniature person! Yes, she is obliged."). It is a matter of nudging one's interlocutor's current neuronal activation-point *out* of the attractor-category that has captured it, and *into* a distinct attractor-category. It is a matter of trying to change the probability, or the robustness, or the proximity to a shared neural prototype-pattern, of one's dialectical opponent's neural behavior.

In the less tractable case where the opponents fail to share a common family of moral prototypes, moral argument must take a different form. I postpone discussion of this deeper form of conflict until the section on moral progress.

7. Moral Virtue

Moral virtues are the various skills of social *perception*, social *reflection*, *imagination*, and *reasoning*, and social *manipulation* that normal social learning produces. In childhood, one must come to appreciate the high-dimensional background structure of social space – its offices, its practices, its prohibitions, its commerce – and one must learn to recognize its local configuration swiftly and reliably. One must also learn to recognize one's own current position within it, and the often quite different positions of others. One must learn to anticipate the normal unfolding of this ongoing commerce, to recognize and help repair its occasional pathologies, and to navigate its fluid structure while avoiding social disasters, both large and small. All this requires skill in divining the social perceptions and personal interests of others, and skill in manipulating and negotiating our collective behavior.

Being skills, moral virtues are inevitably acquired rather slowly, as anyone who has raised children will be familiar. Nor need their continued development ever cease, at least in individuals with the continued opportunities and the intelligence necessary to refine them. The acquired structures within one's neuronal-activation spaces – both perceptual and motor – can continue to be sculpted by ongoing experience and can thus pursue an ever deeper insight into, and an effectively controlling grasp of, one's enclosing social reality. Being skills, they are also differently acquired by distinct individuals, and differentially acquired within a single individual. Each brain is slightly different from every other in its initial physical structure, and each brain's learning history is unique in its myriad details. No two of us are identical in the profile of skills we acquire, which raises our next topic.

8. Moral Character

A person's unique moral character is just the individual profile of his or her perceptual, reflective, and behavioral skills in the social domain. From what has already been said, it will be evident that moral character is distinguished more by its rich diversity across individuals than by its monotony. Given the difficulty in clearly specifying any canonical profile as being uniquely ideal, this is probably a good thing. Beyond the unending complexity of social space, the existence of a diversity of moral characters simply reflects a healthy tendency to explore that space and to explore the most effective styles of navigating it. By this I do not mean to give comfort to moral nihilists. That would be to deny the reality of social learning. What I am underwriting here is the idea that long-term moral learning across the human race is positively served by tolerating a Gaussian distribution of well-informed "experiments" rather than by trying to impose a narrow and impossible orthodoxy.

This view of the assembled moral virtues as a slowly acquired network of skills also contains an implicit critique of a popular piece of romantic nonsense, namely, the idea of the "sudden convert" to morality, as typified by the "tearful face of the repentant sinner" or the postbaptismal "born-again" charismatic Christian. Moral character is not something – is not *remotely* something – that can be acquired in a day by an act of will or by a single major insight.

The idea that it can be so acquired is a falsifying reflection of one or the other of two competing conceptions of moral character, herewith discredited. The first identifies moral character with the acceptance of a canonical set of behavior-guiding rules. The second identifies moral character with a canonical set of desires, such as the desire to maximize the general happiness, and so on. Perhaps one can embrace a set of rules in one cathartic act, and perhaps one can permanently privilege some set of desires by a major act of will. But neither act can result in what is truly needed, namely, an intricate set of finely honed perceptual, reflective, and sociomotor skills. These take several decades to acquire. Epiphanies of moral commitment can mark, at most, the initiation of such a process. Initiations are welcome, of course, but we do not give children a high school diploma simply for showing up for school on the first day of the first grade. For the same reasons, "born-again" moral characters should probably wait a similar period of time before celebrating their moral achievement or pressing their moral authority.

9. Moral Pathology

Moral pathology is a large topic, since, if there are many different ways to succeed in being a morally mature creature, there are even more ways in which one might fail. But as a first pass, moral pathology consists in the partial absence, or subsequent corruption, of the normal constellation of perceptual, reflective, and behavioral skills under discussion. In terms of the cognitive theory that underlies the present approach, it consists in the failure to achieve, or subsequently to activate normally, a suitable hierarchy of moral prototypes. And at the lowest level, this consists in a failure, either early or late, to achieve and maintain the proper configuration of the brain's 10^{14} synaptic weights, the configuration that sustains the desired hierarchy of prototypes and makes possible their appropriate activation.

The terms "normally," "suitable," "proper," and "appropriate" all appear in this quick characterization, and they will all owe their sense to a complex mix of *functional* understanding within cognitive neurobiology and genuine *moral* understanding as brought to bear by common sense and the civil and criminal law. The point here urged is that we can come to understand how displays of moral incompetence, both major and minor, are the reflection of specific functional failures, both large and small, within the brain. This is not a speculative remark. Thanks to the increasing availability of brain-scanning technologies such as Positron Emission Tomography (PET) and Magnetic Resonance Imaging (MRI), neurologists are becoming familiar with a variety of highly specific forms of brain damage that display themselves in signature forms of cognitive failure in moral perception, moral reasoning, and social behavior (Damasio, Tranel, and Damasio 1991, Damasio 1994, Bechara et al. 1994, Adolphs et al. 1996).

Two quick examples will illustrate the point. The neurologists Antonio and Hanna Damasio have a patient, known in the literature as "Boswell," who is independently famous for his inability, because of bilateral lesions to his hippocampus, to lay down any new long-term memories. Since his illness, his "remembered past" is a moving window that reaches back no more than forty seconds. More important for our purposes, it later emerged that he also displays a curious inability to "see evil" in pictures of various emotionally charged and potentially violent scenes. In particular, he is unable to pick up on the various negative emotions as expressed in people's *faces*, and he will blithely confabulate innocent explanations of the socially and morally problematic scenes shown him. (By contrast, he picks up the expression of negative emotions in people's *voices* just fine.) There is nothing wrong with Boswell's eyes, however. His cognitive deficit lies roughly a dozen synaptic steps and a dozen neuronal layers behind his retinas.

As MRI scans revealed, Boswell's herpes-simplex encephalitis also damaged the lower half of both temporal lobes, which includes the area called "IT" (infero-temporal), known for its critical role in discriminating individual human faces and in coding facial expressions. Though Boswell can still recognize the identity of faces well known to him before the illness (movie stars and presidents, for example), his moral perception has been selectively impaired in the manner described.

A second Damasio patient, EVR, had a normal life as a respected accountant, devoted father, and faithful husband. When he was in his mid forties, he had a ventromedial frontal brain tumor successfully removed, and subsequent tests revealed no change in his original IQ of 140. But within six months of the surgery, he lost his job for rampant irresponsibility, made a series of damaging financial decisions, was divorced by his frustrated wife, briefly married and then was left by a prostitute, and generally became incapable of the normal prudence that guides complex planning and intricate social interactions. Subsequent MRI scans confirmed that the surgical removal of the tumor had lesioned the relatively small axonal pathway that connects the ventromedial frontal cortex (the seat of complex planning) to the amygdala (a primitive limbic area that apparently embodies fear, anxiety, and disgust). The functional consequence of this break in the wiring was to *isolate* EVR's practical reasonings from the "visceral" somatic and emotional reactions that normally accompany the rational evaluation of practical alternatives. In normals, those "somatic markers" (as the Damasios have dubbed them) constitute an important dimension of socially relevant information and a key factor in inhibiting one's decisions. In EVR, they have been cut out of the loop, resulting in the sorts of behavior described.

These two failures, of moral perception and moral behavior, respectively, resulted from sudden illness and consequent damage to specific brain areas, which is what brought these patients to the attention of the medical profession and led to their detailed examination. But these and many other neural deficits can also appear slowly, as a result of developmental misadventures and other chronic predations – childhood infections, low-level toxins, abnormal metabolism, abnormal brain chemistry, abnormal nutrition, maternal drug use during pregnancy, and so forth. There is no suggestion, let me emphasize, that all failures of moral character can be put down to structural deficits in the brain. A proper moral education – that is, a long stretch of intricate socialization – remains a necessary condition for acquiring a well-formed moral character, and no doubt the great majority of failures, especially the minor ones, can be put down to sundry inadequacies in that process.

Even so, the educational process is thoroughly entwined with the developmental process and deeply dependent on the existence of normal brain structures to embody the desired matrix of skills. At least some failures of moral character, therefore, and especially the most serious failures, are likely to involve some confounding disability or marginality at the level of brain structure and/or physiological activity. Therefore, if we wish to wisely address such major failures of moral character, in the law and within the correctional system, we would do well to understand the many dimensions of neural failure that can collectively give rise to them. We can't fix what we don't understand.

10. Moral Correction

Consider first the structurally and physiologically *normal* brain whose formative social environment fails to provide a normal moral education. The child's experience may lack the daily examples of normal moral behavior in others, it may lack opportunities to participate in normal social practices, it may fail to see others deal successfully and routinely with their inevitable social conflicts, and it may lack the normal background of elder sibling and parental correction of its perceptions and its behavior. For the problematic young adult that results, moral correction will obviously consist in the attempt somehow to make up for a missed or substandard education.

That can be difficult. The cognitive plasticity and eagerness to imitate found in children is much reduced in a young adult. And a young adult cannot easily find the kind of tolerant community of innocent peers and wise elders that most children are fortunate to grow up in. Thus not one but two important windows of opportunity have been missed.

The problem is compounded by the fact that children in the impoverished social environments described do not simply fail to learn. They may learn quite well, but *what* they learn is a thoroughly twisted set of social and moral prototypes and an accompanying family of skills that – while crudely functional within the impoverished environment that shaped them, perhaps – are positively *dys*functional within the more coherent structure of society at large. This means that the young adult has some substantial *un*learning to do. Given the massive cognitive "inertia" characteristic even of normal humans, this makes the corrective slope even steeper, especially when young adult offenders are incarcerated in a closely knit prison community of similarly twisted social agents.

This essay was not supposed to urge any substantive social or moral policies, but those who do trade in such matters may find relevant the following purely factual issues. America's budget for state and federal prisons is said to be larger than its budget for *all* of higher education – for all of its elite research universities, massive state universities, and myriad liberal arts colleges, technical colleges, and junior colleges combined. It is at least conceivable that our enormous penal-system budget might more wisely be spent on prophylactic policies aimed at raising the quality of the social environment of disadvantaged children rather than on policies that struggle, against much greater odds, to repair the damage once it has been done.

A convulsive shift, of course, is not an option. Whatever else our prisons do or do not do, they keep at least some of the dangerously incompetent social agents and the outright predators off our streets and out of our social commerce. But the plasticity of the young over the old poses a constant invitation to shift our corrective resources childwards as due prudence dictates. This policy suggestion hopes to reduce the absolute input to our correctional institutions. An equally important issue is how, in advance of such a "utopian" breakthrough, to increase the rate at which the prisons are emptied, to which topic I now turn.

A final point, in this regard, about normals. The cognitive plasticity of the young – that is, their unparalled capacity for learning – is owed to neurochemical and physiological factors that fade with age. (The local production and diffusion of nitric oxide within the brain is one theory of how some synaptic connections are made selectively subject to modification, and there are others.) Suppose that we were to learn how to *re-create* in young adults, temporarily and by neuropharmacological means, that perfectly normal regime of neural plasticity and learning aptitude found in children. In conjunction with some more effective programs of resocialization than we currently command (without them, the pharmacology will be a waste of time), this might relaunch the "disadvantaged normals" into something much closer to a normal social trajectory and out of prison for good.

There remain, alas, the genuine abnormals, for whom moral correction is first a matter of trying to repair or compensate for some structural or physiological defect(s) in brain function. Even if these people are hopeless, it will serve social policy to identify them reliably, if only to keep them permanently incarcerated or otherwise out of the social mainstream. But some, at least, will not be hopeless. Where the deficit is biochemical in nature – giving rise to chronically inappropriate emotional profiles, for example – neuropharmacological intervention, in the now-familiar form of chronic subdural implants, perhaps, will return some victims to something like a normal neural economy and a normal emotional profile. That will be benefit enough, but these individuals will then also be candidates for the resocialization techniques imagined earlier for disadvantaged normals.

This discussion presumes far more neurological understanding than we currently possess, and is plainly speculative as a result. But it does serve to illustrate some directions in which we might well wish to move, once our early understanding here has matured. In any case, I shall close this discussion by reemphasizing the universal importance of gradual social-ization by long interaction with a moral order already in place. We will never create moral character by medical intervention alone. There are too many trillions of synaptic connections to be appropriately weighted, and only long experience can hope to do that superlatively intricate job. The whole point of exploring the technologies just mentioned will be to maximize everyone's chances of engaging in and profiting from that traditional and irreplacable process.

11. *Moral Diversity*

In this section I refer not to the high-dimensional bell-curve diversity of moral characters within a given culture at a given time, but to the nonidentity across two cultures, separated in space and/or in time, of the overall *system* of moral prototypes and prized skills common to most normal members of each. Such major differences in moral consciousness typically reflect differences in the two cultures' substantive economic circumstances, in the peculiar threats to social order with which they have to deal, in the technologies they command, in the metaphysical beliefs they happen to hold, and in other accidents of history.

Such diversity, when discovered, is often seen as grounds for a blanket skepticism about the objectivity or reality of moral knowledge. That was certainly its effect on me in my later childhood, a reaction reinforced by the astonishingly low level of moral argument I would regularly hear from my more religious schoolchums, and even from the local pulpits. But that is no longer my reaction, for throughout history there have been even greater differences, between distinct cultures, where *scientific knowledge* is concerned, and comparable blockheadedness in purely "factual" rea-soning (think of "New Age medicine," for example, or of "UFOlogy"). But this diversity and equally lamentable sloppiness does not underwrite a blanket skepticism about the possibility of scientific knowledge. It shows merely that scientific knowledge is not easy to come by, and that its achievement requires a long-term process of careful and honest evalu-ation of a wide variety of complex experiments over a substantial range of human experience. Which raises our next topic.

12. *Moral Progress*

If it exists – there is some dispute about this – moral progress con-sists in the slow change and development, over historical periods, of the moral prototypes we teach our children and impose on derelict adults. This developmental process is gradually instructed by our collective experience of a collective life lived under those perception-shaping and behavior-guiding prototypes.

From the neurocomputational perspective, this process looks different only in its ontological focus – the *social* world as opposed to the *natural* world – from what we are pleased to call *scientific progress*. In the natural sciences as well, achieving adult competence is a matter of acquiring a complex family of perceptual, reflective, and behavioral skills in the relevant field. And there, too, such skills are embodied in an acquired set of structural, dynamical, and manipulational prototypes. The occasional deflationary voice to the contrary, our scientific progress over the centuries is a dramatic reality, and it results from the myriad instructions (often painful) of an ongoing experimental and technological life lived under those same perception-shaping and behavior-guiding scientific prototypes.

Our conceptual development in the moral domain, I suggest, differs only in detail from our development in the scientific domain. We even have institutions whose job it is to continually fine-tune and occasionally reshape our conceptions of proper conduct, permissible practice, and proscribed behavior. Local, state, and federal legislative bodies spring immediately to mind, as does the civil service, and so too do the several levels of the judiciary and their ever-evolving bodies of case-law and decision-guiding legal precedents. As with our institutions for empirical science, these socially focused institutions typically outlive the people who pass through their offices, often by centuries and sometimes by many centuries. And, as with the payoff from our scientific institutions, the payoff here is the accumulation of unprecedented levels of recorded (social) experience, the equilibrating benefits of collective decision making, and the resulting achievement of levels of moral understanding that are unachievable by a single individual in a single lifetime.

To this overarching parallel it may be objected that science addresses the ultimate nature of a fixed, stable, and independent reality, while our social, legislative, and legal institutions address a plastic reality that is deeply dependent on the organizing activity of humans. But this presumptive contrast disappears almost entirely when one sees the acquisition of both scientific and moral wisdom as the acquisition of sets of *skills*. Both address a presumptively *im*plastic part of their respective domains – the basic laws of nature in the former case, and basic human nature in the latter. And both also address a profoundly *plastic* part of their respective domains – the articulation, manipulation, and technological exploitation of the natural world in the case of working science, and the articulation, manipulation, and practical exploitation of human nature in the case of working morals and politics. A prosperous city represents simultaneous success in both dimensions of human cognitive activity. And the resulting artificial technologies, both natural and social, make possible a deeper insight into the basic character of the natural universe, and of human nature, respectively.

13. Moral Unity/Systematicity

This parallel with natural science has a further dimension. Just as progress in science occasionally leads to welcome unifications within our understanding – as when all planetary motions come to be seen as special cases of projectile motion, and all optical phenomena come to be seen as special cases of electromagnetic waves – so also does progress in moral theory bring occasional attempts at conceptual unification – as when our assembled obligations and prohibitions are all presented (by Hobbes) as elements of a *social contract*, or (by Kant) as the local instantiations of a *categorical imperative*, or (by Rawls) as the reflection of *rules rationally chosen from behind a veil of personal ignorance*. These familiar suggestions, and others, are competing attempts to unify and systematize our scattered moral intuitions or antecedent moral understanding, and they bring with them (or hope to bring with them) the same sorts of virtues displayed by intertheoretic reductions in science, namely, greater simplicity in our assembled conceptions, greater consistency in their application, and an enhanced capacity (born of increased generality) for dealing with novel kinds of social and moral problems.

As with earlier aspects of moral cognition, this sort of large-scale cognitive achievement is also comprehensible in neurocomputational terms, and seems to involve the very same sorts of neurodynamical changes that are (presumptively) involved when theoretical insights occur within the natural sciences. Specifically, a wide range of perceptual phenomena – which (let us suppose) used to activate a large handful of distinct moral prototypes, m_1, m_2, m_3, ..., m_n – come to be processed under a new regime of recurrent manipulation (recall the recurrent neuronal pathways of Figure 3.1*b*) that results in their all activating an unexpected moral prototype M, a prototype whose typical deployment has hitherto been in other perceptual domains entirely, a prototype that now emerges as a *superordinate* prototype of which the scattered lesser prototypes, m_1, m_2, m_3, ..., m_n, can now be seen, retrospectively, as so many *subordinate* instances.

The preceding is a neural-network description of what happens when, for example, our scattered knowledge in some area gets *axiomatized*. But axiomatization, in the linguaformal guise typically displayed in textbooks, is but one minor instance of this much more general process, a process that embraces the many forms of *non*discursive knowledge as well, a process that embraces science and ethics alike.

14. *Reflections on Some Recent "Virtue Ethics"*

As most philosophers will perceive, the general portrait of moral knowledge that emerges from neural-network models of cognition is a portrait already under active examination within moral philosophy, quite independent of any connections it may have with cognitive neurobiology. Its original champion is Aristotle, and its current research community includes figures as intellectually diverse as Mark Johnson (1993), Owen Flanagan (1991), and Alasdair MacIntyre (1981), all of whom came to this general perspective for reasons entirely their own. For the many reasons outlined in this paper, I am compelled to count myself among them. But I am not entirely comfortable in this group, for two of the philosophers just mentioned take a view very different from mine on the matter of moral progress. Flanagan (1996) has expressed doubts that human moral consciousness ever makes much genuine "progress," and he suggests that its occasional changes are better seen as just a directionless meander made in local response to our changing economic and social environment.

MacIntyre (1981) voices a different but comparably skeptical view, wherein he hankers after the lost innocence of pre-Enlightenment human communities, which were much more tightly knit by a close fabric of shared social practices, which practices provided the sort of highly interactive and mutually dependent environment needed for the many moral virtues to develop and flourish. He positively laments the emergence of the post-Enlightenment, liberal, secular, and comparatively anonymous and independent social lives led by modern industrial humans, since the rich soil necessary for moral learning, he says, has thereby been impoverished. The familiar moral virtues must now be acquired, polished, and exercised in what is, comparatively, a social vacuum. If anything, in the last few centuries we have suffered a moral *re*gress.

I disagree with both authors, and will close by outlining why. I begin with MacIntyre, and I begin by conceding his critique of the (British)

Enlightenment's cartoonlike conception of *homo economicus,* a hedonic calculator almost completely free of any interest in or resources for evaluating the very desires that drive his calculations. I likewise concede MacIntyre's critique of the (Continental) Enlightenment's conception of *pure reason* as the key to identifying a unique set of behavior-guiding rules. And my concessions here are not reluctant. I agree wholeheartedly, with MacIntyre, that neither conception throws much light on the nature of moral virtue.

But as crude as these moral or metamoral ideas were, they were still a step up from the even more cartoonlike conceptions of *homo sheepicus* and *homo infanticus* relentlessly advanced by the pre-Enlightenment Christian church. Portraying humanity as sheep guided by a supernatural Shepherd, or as children beholden to a supernatural Father, was an even darker self-deception and was an even less likely way to lead humans up the ladder of moral understanding.

I could be wrong in this blunt assessment, and if I am, so be it. For the claim of the preceding paragraph does *not* embody the truly important argument for moral progress at the hands of the Enlightenment. That argument lies elsewhere. It lies in the permanent opening of a tradition of cautious *tolerance* for a diversity of local communities, each bonded by its own fabric of social practices; it lies in the establishment of lasting institutions for the principled *evaluation* of diverse modes of social organization, and for the institutionalized *criticism* of some and the systematic *emulation* of others. It lies, in sum, in the fact that the Enlightenment broke the hold of a calcified moral dictatorship and replaced it with a tradition that was finally prepared to learn from its deliberately broad experience and its inevitable mistakes in first-order moral policy.

Once again, I am appealing to a salient parallel. The virtue of the Enlightenment, in the moral sphere, was precisely the same virtue as that displayed in the scientific sphere, namely, the legitimation of responsible theoretical diversity and the establishment of lasting institutions for its critical evaluation and positive exploitation. It is this long-term process, rather than any particular moral theory that might fleetingly engage its attention, that marks the primary achievement of the Enlightenment.

MacIntyre begins his introduction with a thought-provoking science-fiction scenario about the loss of an intricate practical tradition that alone gives life to its corresponding family of theoretical terms and the relative barrenness of the terms' continued use in the absence of that sustaining tradition. This scenario embodies the essentials of his critique of our moral history since the Enlightenment. But we can easily construct a parallel critique of our *scientific* history since the same period, and that parallel, I suggest, throws welcome light on MacIntyre's peculiar perspective.

Consider the heyday of Aristotelian science, from the fourth century B.C. to the seventeenth century A.D. (even longer than the Christian domination of the moral sphere), and consider the close-knit and unifying set of intellectual and technological practices that it sustained. There is the medical tradition running from Rome's Galen to the four Humors of the late-medieval doctors. There is the astronomical/astrological tradition that extends through Alexandria's Ptolemy to Prague's Johannes Kepler, who was still casting horoscopes for the wealthy despite his apostate theorizing. There is the intricate set of industrial practices maintained by the alchemists, from Alexandrian Greeks to seventeenth-century Europeans, which tradition simply owned the vital practices of metallurgy and metal-working, and of dye-making and medicinal manufacture as well. These three traditions, and others that space bids me pass over, were closely linked by daily practice as well as by conceptual ancestry, and they formed a consistent and coherent environment in which the practical and technological virtues of antiquity could flourish. As they did. MacIntyre's first condition is met.

So is his second, for this close-knit "paradise" is well and truly lost, having been displaced by a hornet's nest of distinct sciences, sciences as diverse as astrophysics, molecular biology, anthropology, electrical engineering, solid-state physics, immunology, and thermodynamic meteorology. Modern science now addresses and advances on so many fronts that the research practice of individual scientists and the technological practice of individual engineers is increasingly isolated from all but the most immediate members of their local cognitive communities. And the cognitive virtues they display are similarly fragmented. They may even find it difficult to talk to each other.

You see where I am going. There may well be problems arising from the unprecedented flourishing of the many modern sciences, but losing an earlier and somehow more healthy "golden age" is certainly not one of them. Though real, those problems are simply the price that humanity pays for growing up, and we already attempt to address them by way of interdisciplinary curricula, conferences, and anthologies, and by the never-ending search for explanatory unifications and intertheoretic reductions.

I propose, for MacIntyre's reflection, a parallel claim for our moral, political, and legal institutions since the Enlightenment. Undoubtedly there are problems emerging from the unprecedented flourishing of the many modern industrial societies and their subsocieties, but losing touch with a prior golden age is not obviously one of them. The very real problems posed by moral and political diversity are simply the price that humanity pays for growing up. And as in the case of the scattered sciences, we already attempt to address them by constant legislative tinkering, by the reality-driven evolution of precedents in the judicial record, by toleration of the occasional political "divorce" (e.g., Bosnia, the Soviet Union, the Scottish Parliament), and by the never-ending search for legal, political, and economic unification. Next to the discovery of fire and the polydoctrinal example of ancient Greece, the Enlightenment may be the best thing that ever happened to us.

The doctrinal analog of communitarianism in moral theory is a hyperbolic form of the conservatism of Thomas Kuhn in the philosophy of science, a conservativism that values the (very real) virtues of any given "normal science" tradition (such as Ptolemaic astronomy, classical thermodynamics, or Newtonian mechanics) over the comparatively fragile institutions of collective evaluation, comparison, and criticism that might slowly force their hidden vices into the sunlight and pave the way for their rightful overthrow at the hands of even more promising modes of cognitive organization. One can certainly see Kuhn's basic "communitarian" point: stable practices make many valuable things possible. But tolerant institutions for the evaluation and modification of those practices make even *more* valuable things possible – most obviously, new and more stable practices.

This particular defense of the Enlightenment also lays the foundation for my response to Flanagan's quite different form of skepticism. As I view matters from the neural-network perspective explained earlier in this essay, I can find no difference in the presumptive brain mechanisms and cognitive processes that underwite moral cognition and scientific cognition. Nor can I find any significant differences in the respective social institutions that administer our unfolding scientific and moral consciousness respectively. In both cases, learning from experience is the perfectly normal outcome of both the neural and the social machinery. That means that moral progress is no less possible and no less likely than scientific progress. And since none of us, at this moment, is being shown the instruments of torture in the Vatican's basement, I suggest it is actual as well.

There remains the residual issue of whether the *sciences* make genuine progress, but that issue I leave for another time. The take-home claims of the present essay are that (1) whatever their ultimate status, moral and scientific cognition are on an *equal* footing, since they use the same neural mechanisms, show the same dynamical profile, and respond in both the short and the long term to similar empirical pressures; and (2) in both moral and scientific learning, the fundamental cognitive achievement is the acquisition of *skills*, as embodied in the finely tuned configuration of the brain's 10^{14} synaptic connections.

4

Rules, Know-How, and the Future of Moral Cognition

Professor Andy Clark's splendid essay[1] represents a step forward from which there should be no retreat. Our de facto moral cognition involves a complex and evolving interplay between, on the one hand, the *non*discursive cognitive mechanisms of the biological brain, and on the other, the often highly discursive extrapersonal "scaffolding" that structures the social world in which our brains are normally situated, a world that has been, to a large extent, created by our own moral and political activity. That interplay extends the reach and elevates the quality of the original nondiscursive cognition, and thus any adequate account of moral cognition must address both of these contributing dimensions. An account that focuses only on brain mechanisms will be missing something vital.

I endorse these claims, so compellingly argued by Clark, for much the same reasons that I also endorse the following claims. Our de facto *scientific* cognition involves a complex and evolving interplay between, on the one hand, the *non*discursive cognitive mechanisms of the biological brain, and on the other, the often highly discursive extrapersonal "scaffolding" that structures the social-scientific world in which the brains of scientists are normally situated, a technologically and institutionally intricate world that has been, to a large extent, created by our own scientific activities. That interplay extends the reach and elevates the quality of the original nondiscursive cognition, and thus any adequate account of

---

[1] Andy Clark, "Word and Action: Reconciling Rules and Know-How in Moral Cognition," in R. Campbell and B. Hunter, eds., *Moral Epistemology Naturalized. Canadian Journal of Philosophy*, suppl. vol. 26: 267–90.

scientific cognition must address both of these contributing dimensions. An account that focuses only on brain mechanisms will be missing something vital.

I draw this parallel for many reasons, as will emerge, but a salient reason is that, whatever theoretical story we decide to tell about "situated" cognition, it must meet the experimental test of, not one, but at least *two* important domains of human cognitive activity. A second reason is to emphasize that Clark's (entirely genuine) insights about the "situated" character of our moral cognition do nothing to distinguish it, in any fundamental way, from human cognition in general, including our scientific cognition. And a third reason is that each of these two cognitive domains – the broadly scientific, and the broadly moral – may have a good deal to teach us about the other, once we appreciate that, and how, they are brothers under the skin.

I. The Role of Discursive Rules

While Clark finds an important role for discursive moral rules, within the context of the nondiscursive, connectionist, prototype-centered account of moral knowledge, we must be mindful that the role he finds is profoundly different from the role that tradition has always assumed moral rules to play. I do not mean to suggest that Clark is under any illusions on this score, but many of his readers will be, and so it is appropriate to begin by emphasizing the novelties that we here confront. Clark's story on moral cognition is in no way a critique or a rejection of the recent nondiscursive neural-network models of human and animal cognition.[2] Rather, it is an important and appropriate *augmentation* of that approach. It is a local reflection of his views on situated cognition in general, as outlined in his 1997 book.[3] That more general view is interesting because it finds

---

[2] For a quick and accessible introduction, see P. M. Churchland, *The Engine of Reason, the Seat of the Soul: A Philosophical Journey into the Brain* (Cambridge, MA: MIT Press, 1995). For a sketch of its applications to moral theory in particular, see P. M. Churchland, "Toward a Cognitive Neurobiology of the Moral Virtues," *Topoi* 17 (1998): 83–96. For a more thorough and more neurophysiologically focused introduction, see P. S. Churchland and T. J. Sejnowski, *The Computational Brain* (Cambridge, MA: MIT Press, 1992). For a more philosophically oriented introduction, see P. M. Churchland, *A Neurocomputational Perspective: The Nature of Mind and the Structure of Science* (Cambridge, MA: MIT Press, 1989). For a rigorous mathematical introduction, see R. Rojas, *Neural Networks: A Systematic Introduction* (New York: Springer-Verlag, 1996). The bibliography of any of these works will lead you stepwise into the larger literature.

[3] Andy Clark, *Being There: Putting Brain, Body, and World Together Again* (Cambridge, MA: MIT Press, 1997).

a significant portion of the machinery available to cognition, and a significant portion of the activity of cognition, to lie *outside* the brain. It lies in the extrapersonal public space of drawn diagrams, written arithmetic calculations, spoken and printed arguments, tools of measurement and manipulation, and extranumary "cognitive prosthetics" of many other kinds as well. The idea is that the brain learns to "off-load" certain aspects of some needed computational activity into some appropriate external medium of representation and manipulation, because the job can there be done more easily, quickly, or reliably than inside the brain. Deploying the familiar grade-school recursive procedures ("write down the 6, carry the 1") with pencil on paper, to compute large arithmetical sums, would be a prototypical instance of the "off-loading" phenomenon Clark has in mind, and you can easily begin to generalize from this mundane example. In particular, you can begin to see a cognate role for the linguistic machinery of moral conversation, moral argument, and moral directives.

Now this externalist vision, I believe, is the *right* way to see the role of discursive representations. But it is vital to appreciate that it involves a major shift away from the avowedly *internalist* perspective that dominates traditional moral theory of almost every stripe. According to that tradition, to be moral is to have embraced, accepted, or otherwise internalized a specific set of behavior-guiding rules, which stored rules are then deployed in appropriate circumstances as a salient part of the internal cognitive mechanisms that actually produce intentional behavior. (Once these assumptions are in place, the principal philosophical questions are then pretty much fixed: which of the many possible rules are the truly correct or morally binding rules? And what metaphysical, apodeictic, or empirical circumstance – e.g., God's command, a social contract, pure reason, utility maximization, maxi-min choice from behind a veil, and so on – bestows that vaulted status upon them?) What goes unnoticed in this highly general perspective on moral philosophy, at least until recently, is that it surreptitiously presupposes a background theory about the nature of cognition, a theory that we now have overwhelming reason to believe is empirically false, a theory for which we already possess the outlines of a neuronally based and mathematically embodied alternative, specifically, the vector-coding, matrix-processing, prototype-activating, synapse-adjusting account held out by cognitive neurobiology and connectionist AI.

What changes does this new cognitive perspective require? Several. First and foremost, it requires us to give up the idea that our internal representations and cognitive activities are essentially just hidden, silent versions of the external statements, arguments, dialogues, and chains of reasoning that appear in our overt speech and print. That conception is an old and venerable one, to be sure, for it is the constituting assumption of our dear beloved "folk psychology." And it is also a *natural* one, for, how *else* should we conceive of our inner activities, save on the model of outer speech, our original and (until recently) our only empirical example of a representational/computational system?[4] How else indeed?

But in fact there are other ways, and ignorance of them has been our excuse for far too long. Nonlinguistic creatures (i.e., most of the creatures on the planet) provide the initial motivating cognitive examples. For it is not plausible to portray them as using the same discursive, linguaformal thought processes that we so routinely ascribe to ourselves. After all, why conceive of all animal cognition on the model of an isolated discursive skill that is utterly unique to a single species? But neither is it plausible to dismiss all nonhuman animals as thoughtless, stimulus–response driven brutes. They are far too clever for that. Plainly, we need a third approach, free from a procrustean anthropocentric romanticism on the one hand, and from the dismissive deflation of animal cognitive powers on the other.

II. A Nondiscursive Conception of Cognition

When, in a comparative spirit, we examine the *brains* of terrestrial creatures – their large-scale anatomies, their filamentary microstructures, and their physiological and electrochemical activities – we find a striking *conservation* of form, structure, and function across all vertebrate animals, and especially across the higher mammals, and most especially across the primates, humans included. The basic machinery of cognition is the same in all of us, and it has nothing to do with the structure of declarative sentences, with the rule-governed drawing of inferences from one sentence

---

4 The reader will here recognize Wilfrid Sellars's well-known account of the origins and nature of our folk psychology, as outlined in the closing sections of his classic paper "Empiricism and the Philosophy of Mind," chap. 3 of *Science, Perception, and Reality* (London: Routledge, 1963). Ironically (from our present perspective), Sellars was blissfully convinced that folk psychology was an *accurate* portrayal of our inner cognitive activities. (I recall finding it advisable to downplay my own nascent eliminativism during my dissertation defense, a meeting chaired by that worthy philosopher.) But Sellars's conviction on this point notwithstanding, folk psychology had invited systematic skepticism long before the present, and for reasons above and beyond the recent flourishing of cognitive neurobiology. See, for example, P. M. Churchland, "Eliminative Materialism and the Propositional Attitudes," *Journal of Philosophy* 78, no. 2 (1981): 67–90, now more than twenty years old.

to another, or with the storage and deployment of rules of any kind. Instead, that machinery is wonderfully designed by evolution to subserve the acquisition and deployment of a panoply of *skills* and *abilities*.

Those skills include, most obviously, a broad range of *perceptual* skills, for a creature must learn to discriminate not only colors and shapes, but to recognize such things as the peculiar locomotor gaits of its typical predators and typical prey; the entreaty or hostility in the facial expression of a conspecific; the gathering weariness of an infant, or an adversary; the existence and profile of kin relations and social alliances within one's group; the opportunities to forge and share in such alliances; and the appropriate occasions to express the commitments – such as defense, comfort, and sharing – that go with those alliances. Perception, plainly, can involve considerable conceptual sophistication.

No less important are the *motor* skills that must be acquired. A creature must learn to walk, to run, to climb, or to fly, and so forth. But it must also learn to chase its prey, to groom its conspecifics, to fend off an attack, to make a nest or burrow, to assemble an electric motor, or, if one is an administrator, to do such things as take a company public, or launch the Allied invasion of Normandy. Motor skills, like perceptual skills, can also involve a high degree of conceptual sophistication.

Finally, and not to be sharply separated from the skills already discussed, are the various skills of sensorimotor *coordination* – the skills of matching one's behavior to one's current perceptions, or of using one's ongoing perceptions to steer and modulate one's ongoing behavior. Importantly, much of one's perception involves the recognition of prototypical *processes* that unfold in time, such as falling bodies, flying insects, swimming fish, and fleeing mice. Moreover, the perceptual recognition of such processes consists in the activation of a previously learned prototypical *sequence* of activation-patterns in the relevant neuron population. Accordingly, a creature with sensorimotor coordination can *anticipate* the unfolding of its perceptual environment, for at least a few fractions of a second into the future, and then steer its motor behavior to suit that anticipated environment. It can dodge the falling body, swat the flying insect, and catch the moving fish or mouse. In this basic capacity for sensorimotor coordination lie the origins of all intelligence, and one obvious measure of the degree of intelligence that any creature has achieved is *how far* into the future and across *what range* of phenomena it is capable of projecting the behavior of its environment, and thus how far in advance of that future it can begin to execute appropriately exploitative and manipulative motor behavior. What distinguishes the intelligence of humans from that of all other creatures is not some cognitive discontinuity such as the possession of language. More likely it is our preeminent talent in something we share with all cognitive creatures: we can see further into the future, and can execute motor behavior to exploit that future, than any other creature on the planet

To complete this thumbnail sketch of the basic and nondiscursive cognitive activities common to all terrestrial creatures, suppose now that many species of animal acquire the ability to play and replay "off-line" (i.e., in some fashion that disconnects them from their normal motor sequelae) the various prototypical sequences of activation patterns – both perceptual and motor – that prior experience of the world has taught them. The reader will recognize these activational excursions as instances of daydreaming or projective imagination. As launched in specific perceptual circumstances, they will constitute episodes of "vicarious exploration" of the environment. That is, they will constitute episodes of subjunctive and practical *reasoning*. We are here contemplating a conception of high-level cognitive activity that is recognizably true of ourselves, but which contains no hint of discursive representations and rule-governed activity. This basic conception is all the more interesting because an explanatorily fertile theory of its general nature (i.e., the vector-processing story of connectionism) is already in place, and because that abstract functional theory coheres very nicely with the implementation-level story of neurons and synapses provided by the empirical neurosciences. Indeed, it was our study of the latter that originally inspired our development of the former.

III. Moral Cognition and the Novelty of Rules

"Oh, very well," one might reply, a tad impatiently, "so a nondiscursive form of cognition underlies all of the more advanced forms; but don't we leave that original and primitive form behind when we enter the domain of morality and complex social cognition?"

Not at all. We can see this vital fact immediately by looking at all of the other social mammals on the planet – baboon troops, wolf packs, dolphin schools, chimpanzee groups, lion prides, and so on – and by observing in them the same complex ebb and flow of thoughtful sharing, mutual defense, fair competition, familial sacrifice, staunch alliance, minor deception, major treachery, and the occasional outright ostracism that we see displayed in human societies. [5] Most important for the present issue, none of these other instances of complex social order possesses a language, or any other form of external "cognitive scaffolding," on which to "off-load" some of their social/moral cognition. Their social cognition is conducted *entirely* within the more primitive and nondiscursive form of cognition we have here been discussing. And so, quite evidently, is the greater part of social cognition in human society as well. Typically, it is only when something goes *wrong* with our well-oiled social interactions that we bring into play the discursive scaffolding of rules and moral argument and laws and court procedures.

Even when that external machinery does get deployed, it is the original and more basic form of cognition that does the deploying. Rules are useless unless the capacity for reliable *perception* of their categories is already in place, and such perception depends utterly on the inarticulable processes of vector coding and prototype activation. Moreover, as neural-network models have taught us, a perceptually competent network embodies a great deal of *knowledge* about the general structure of its perceptual environment, knowledge that is embodied in the configuration of its myriad (in humans, 10^{14}) synaptic connections, knowledge that is largely or entirely *inarticulable* by its possessor. There is no hope, to repeat the point, that we can capture the true substance of any human's moral knowledge by citing some family of "rules" that he or she is supposed to "follow," nor is there any hope of evaluating that person's character by evaluating the specific rules within any such internalized family. At the level of individual human cognition, it simply doesn't work that way.

I have pressed this point, perhaps over-pressed it, partly because I wish to uproot an almost universal misconception about the nature of human moral cognition, but also, and correlatively, because I wish to emphasize the genuine *novelty* represented by the evolutionary emergence of language and the cultural emergence of discursive rules. Their emergence makes an enormous difference to the character and quality of our collective moral life. They constitute, as C. A. Hooker would put it,[6] and

---

[5] Appeals to ethology are not always welcome in moral philosophy, but we had better get used to them. The traditionally unquestioned gap between "rational man" and "the unreasoning brutes" is no more substantial than is the division, so long revered in ancient cosmology, between the "sublunary realm" and the "superlunary realm." For a recent and exemplary exploration of what the animal kingdom may have to teach us about the nature of morality, see A. MacIntyre, *Dependent Rational Animals: Why Human Beings Need the Virtues* (La Salle IL: The Open Court, 1999).

[6] C. A. Hooker, *Reason, Regulation, and Realism: Toward a Regulatory Systems Theory of Reason and Evolutionary Epistemology* (Albany, NY: SUNY Press, 1995). This provocative book

Clark would surely agree, a *new level of regulative machinery* to help shape the conduct of our collective affairs, a kind of machinery that had never existed before. They provide us with something the other social animals still do not have. First, they provide a medium for the accumulation of useful social doctrine over periods far in excess of an individual human's lifetime. Second, they provide a system for the collective discussion and local application of that (presumptive) practical wisdom. And third, they enable procedures, consistent across time and circumstance, for identifying and penalizing violations of the discursive rules that (partly) embody that wisdom. They do *not* bring moral reasoning into existence for the first time, and they do *not* provide a conceptual model remotely adequate to the phenomenon of moral cognition in single individuals and nonhuman animals, but they *do* change our lives profoundly.

In fact, as I shall now turn to argue, they change our lives even more profoundly than Clark has urged, and they hold the potential to *further* transform human life, to a degree and in dimensions that his own discussion does not begin to suggest. Specifically, I believe that Clark's own position concerning the importance of extracortical cognitive scaffolding holds the key to understanding how human moral *progress* is not only possible and actual, but still lies mostly ahead of us.

Let me approach these claims by looking at the sorts of rule-based regulative machinery displayed in ancient but postcursive societies. The Judeo-Christian Old Testament provides a roughly typical example: a handful or two of rules, plus a tradition of rabbis, priests, or village elders to officiate their application and enforcement.

In this case, the rules are the now-curious Ten Commandments, plus some now highly uncomfortable Regulations on matters such as the "proper" administration of slavery and indentured servitude (for example, it's OK to beat slaves senseless, as long as you don't actually kill them, Exodus 21:20), on the proper treatment of witches (they must be put to death, Exodus 22:18), and on the proper respect for parents (anyone who curses – curses! – his mother or father must be put to death, Exodus 21:17). Collectively, this body of social legislation, from Exodus 20:1 to 23:31, looks less like the divinely delivered distillation of moral excellence it purports to be, and more like a clumsy attempt, by a profoundly poor and primitive people, to maintain social cohesion against competing human societies, to maintain a minimum of social order within the

presents a general theory of the nested hierarchy of regulatory mechanisms that biological, social, and intellectual evolution have progressively assembled on this planet.

preferred group, and to achieve both aims by instilling stark terror, both metaphysical (the Jealous God) and temporal (prompt execution), into the hearts of the people to be controlled. This Covenant with God is sealed by His promising, in return for our coerced faith, divine intervention in and support for the gradual takeover of all neighboring nations and the subsequent geographical expulsion of the "alien" peoples that constitute them (Exodus 23:20–31). (Whatever happened, one wonders, to the Tenth Commandment, only just laid down – the one that precludes coveting thy neighbor's house or other belongings?)

Contradictions aside, this body of legislation is curious for a number of reasons; first, for the positive law that it contains. Some requirements now appear just silly, such as the practice of regularly sacrificing goats and young bulls as mandatory gestures of solidarity with Jehovah. Other laws are decidedly darker, as with "Thou shalt not suffer a witch to live" (Exodus 22:18). A law requiring such harsh treatment for nonexistent things seems a needless and foolish luxury, at best, and a palpable cruelty if, at worst, the category was intended to include those women – who claim to hear spirit voices and who engage in opaque practices – whom we moderns would now identify as the innocent victims of schizophrenia, a morally neutral brain disorder. The New International Version of the Bible attempts to finesse this embarrassing probability by offering the alternative translation, "Do not allow a *sorceress* to live." Unfortunately, with sorceresses also being nonexistent, this leaves the original puzzle about divine laws for empty categories untouched, and it prompts the further question, "A sorcer*er* is OK?"

This legislative corpus is further curious for the laws that it does *not* contain. For example, there is neither Commandment nor Regulation concerning the proper care and treatment of *children*. It is hard to imagine a more fundamental need for any society, or a more compelling moral imperative for any adult, than the protection and rearing of the children of one's community. (Even baboon troops are faithful at doing that.) And yet this ancient legislative corpus, allegedly divine in its provenance, is simply silent on the matter.

Withal, and despite their primitive character, such ancient bodies of extracortical cognitive scaffolding surely helped to sustain a much more cohesive, effective, and productive social order than could ever have been achieved in their absence. I have no desire to minimize *that* contrast. It is enormous. But my principal aim in pointing out some of the more obviously benighted aspects of the Old Testament's social legislation is to highlight a *second* contrast, one of comparable magnitude and importance.

Specifically, I ask you to compare the crude and tiny body of extracortical social-cognitive scaffolding found in the legal and economic strictures of Exodus to the vast and well-tuned body of social-cognitive scaffolding found in the legal and economic systems of a modern country such as England, France, Canada, or the United States.

IV. The Contrast between Ancient and Modern Scaffolding

A body of behavior-controlling legislation adequate to run an agrarian, bronze-age village is not remotely adequate to run a modern industrial nation with its tens of millions of people and its complex, trillion-dollar, high-tech economy. Our legislation must address practices and facilitate activities of which ancient peoples had little or no conception. The regulation of large corporations, of labor unions, of the stock market, of the nation's banks and interest rates, of agricultural and environmental policy, of pharmaceutical testing and prescription policy, of school curriculums and scientific research policy, of hospitals and penitentiaries, of intellectual property and its industrial applications, of court procedures at the local, state, and national levels, of traffic behavior on our streets and highways, of licensing for electrical contractors, airline pilots, pharmacists, and a thousand other novel professions – these are all matters whose regulation is essential to the health and well-being of modern society, but whose existence went unanticipated by ancient peoples.

The point is not just that we moderns have accumulated more things to regulate than the ancients, although that is certainly true. The important point is that most of these novel phenomena were *created*, partly or wholly, by the initiation of new practices governed by new regulations. There would be no corporations, stock markets, banks, universities, or supreme courts but for the various sorts of carefully regulated human practices that make them possible. The extracortical cognitive scaffolding to which Clark has so aptly drawn our attention is now a glittering skyscraper of monumental proportions. It makes the ancient but cognate scaffolding of Exodus look like a plaster hut by comparison. We have constituted ourselves into a Leviathan that even Hobbes could not have anticipated.

This contrast, I assert, represents substantial moral progress on the part of the human race. Of the matters addressed by ancient legislation, we have simply put some aside entirely, and we regulate the others far more consistently, systematically, sensitively, and wisely than did the ancients. This much is unsurprising, perhaps. We have the advantage of more than two millennia of additional social experience, and we now have the luxury of well-tuned social machinery, with long institutional memories, devoted to the case-by-case administration of our more deeply informed discursive legislation.

This, however, is but a small part of the progress to which we can rightly lay claim. More important still is the expanding universe of new *kinds* of social practices, practices brought into existence by the continued development of new sorts of cognitive scaffolding and new topics of discursive legislation. A primitive villager in the Levant could aspire to many things, perhaps, but he or she could not aspire to be a securities investigator, a labor lawyer, a real estate agent, a software engineer, a congressional lobbyist, a child psychologist, a macroeconomist, a newspaper columnist, a law professor, or a researcher into the genetic basis of various diseases. All of these regulated activities, and a thousand others here unmentioned, constitute new contributions to the well-being of humankind and new dimensions of activity in which people can display excellence, mediocrity, or failure. The high-dimensional web of mutual dependence that now embraces each of us delivers a panoply of goods and services, provides many layers of personal protection to each of us, and affords endless opportunities for self-realization, most of a kind that never existed before.

It may be objected that, even where it is realized, the progress here celebrated is more a matter of our having upgraded the quality and the vitality of the social ocean in which all of us swim, than it is a matter of our having upgraded the personal moral virtues of the average individual human beings who happen to swim in it. With this claim, regrettably, I must largely agree. While the procedural and legislative virtues that constitute a modern nation like Canada or the United States no doubt "rub off" to some degree on its individual people – if only by way of the high standards of the examples it continually sets – the moral character of an average modern North American is probably little superior to the moral character of an average inhabitant of the ancient Levant. The bulk of our moral progress, no doubt, lies in our collective institutions rather than in our individual hearts and minds.

A relevant parallel here concerns our *scientific* progress, which has also transformed our world. Here also, the bulk of our progress resides primarily in our collective institutions of research, education, and technology. Some of that accumulated wisdom clearly "rubs off" on the minds of individual humans, if only because the professions they assume often require some expertise in some smallish area of scientific or technological skill. But on the whole, the scientific understanding of an average modern

North American is probably little superior to the overall scientific understanding of one of Moses' contemporaries.

Little superior, but still *somewhat* superior. And small increments are precious because they can yield large differences in the collective quality of life, especially when those marginally improved individual social and intellectual virtues are exercised in an institutional environment that is itself the repository of much accumulated wisdom. This is as true, and as important, in the moral sphere as it is in the scientific sphere. As I remarked in my opening paragraph, the interplay between the personal and the extrapersonal levels extends the reach and elevates the quality of the individual's original nondiscursive cognitive activities. Plainly, I assert, there has been real progress here, at both levels of cognition, and in both the scientific and the moral domains. And the dynamic of that progress is much the same in both domains: we *learn* from our unfolding *experience* of a world that is partly *constructed* by our own activities.

V. On the Requirements for Future Moral Progress

You see, once more, where I am going: if we can come this far, why not go farther still? Specifically, if the introduction of extracortical cognitive scaffolding gives humans a "leg up" in some cognitive domains, and if the articulation and improvement of that scaffolding, over time and accumulated experience, leads to further improvements in the quality of our cognition in that domain, then why should we not aspire to make *further* improvements in the character and content of our current extracortical scaffolding, so as to make yet further advances in the quality of the cognition at issue?

We may look, once more, at the history of our *scientific* progress for possible insights on how this might unfold in the moral domain. What sorts of things distinguish modern science from the science of the Egyptians and the Babylonians? Most obviously, we have acquired, in sequence, such things as systematic geometry, the algebra of arithmetic unknowns, modern analytic geometry, the infinitesimal calculus, and modern computational theory. Equally obviously, we have escaped the ancient conceptual frameworks of geocentrism, of earth, air, fire, and water, and of "folk physics" generally. Our extrapersonal scaffolding now deploys a new framework of concepts, more penetrating than the old, and more reflective of the world's real makeup.

The social domain shows *some* of the same sorts of advances. We do use modern mathematics to serve the making of economic policy (think of the Federal Reserve Board and its macroeconomic models), and to sustain the nation's monetary activities on a minute-by-minute basis (think of the e-network and the computational facilities that underlie your use of a credit card at the supermarket checkout counter). As well, our conceptions of proper social behavior have certainly changed. (For example, Exodus prohibits the charging of interest on loans, but modern industrial society would collapse without that crucial practice.) On the whole, however, our self-conception and our social technologies show little of the truly radical change evident in our modern scientific conception of the purely natural world.

This is because, I suggest, the neurobiological, cognitive, and social sciences have yet to achieve the major conceptual advances achieved in physics, chemistry, and biology. Bluntly, the cognitive scaffolding that sustains our social lives is still laboring under the burden of a comparatively primitive conceptual framework. "Folk *physics*" may be gone from our enveloping institutions, but "folk *psychology*" is still very much with us, at least in our social institutions.

My point here is not to trash folk psychology: it performs yeoman service for us, and will continue to do so for some time to come. My point is rather that a still deeper conception of the springs and wheels of human nature might perform all of those same services, and many new ones besides, even better than does our current conception.

The geocentric astronomy of Aristotle, Hipparchus, and Ptolemy – to cite a relevant parallel – allowed us to predict the motions of the planets with some precision, and it allowed us to navigate all of the Earth's oceans without getting lost by more than a few hundred kilometers. But in other respects, it was a conceptual and technological straitjacket that simply had to be shed if we were to understand the heavens in general. And when it finally was displaced, the door opened for such novelties as geosynchronous communication satellites and hand-held GPS (Global Positioning System) devices that will fix one's current position on Earth's surface to within a meter. That technology, and a hundred others, are now an integral part of our personal and institutional activities: they have been absorbed by, and are transforming, the extrapersonal cognitive scaffolding that structures our lives.

Similarly, I suggest, will the continuing development of sciences such as cognitive neuroscience, social psychology, neuropathology, neuropharmacology, and vector algebra (the mathematics of neural nets) eventually become absorbed into the extrapersonal, social-level scaffolding that already structures our interpersonal lives. And by being absorbed, it will change that scaffolding, and with it, our moral practices and our moral conceptions. It will afford us the opportunity to hone entirely new nondiscursive cognitive skills, as we learn to navigate a social environment containing novel structures and novel modes of interaction. It will permit a deeper insight into the intricate dance that is each person's unfolding consciousness and thus make possible a deeper level of mutual understanding, care, and protection. It will reconfigure our legal practices, our correctional practices, our educational practices, and perhaps even our recreational and romantic practices.

Clark's skepticism here notwithstanding, the moral domain evidently offers as much prospect for radical progress as does any other domain of cognitive activity. And such progress will be achieved not because – in a runaway spirit of mad-dog reductionism – we *turn our backs* on the social-level cognitive machinery. On the contrary. The current office*holder* may be tossed out on its ear, but the high-level *office* will remain. It will then be occupied, however, by a system of concepts and an accompanying vocabulary grounded in a more deeply informed and technologically more powerful theory of human nature. It will then do all of the old jobs better – those that are worth doing, anyway – and endless new jobs to boot. Accordingly, now is hardly the time to become faint of heart or feeble of vision. The relevant sciences are pregnant with promise, and their effects on social practice are already being felt. The virtues of extrapersonal cognitive scaffolding remain obvious, to be sure. But it is equally obvious that new and better scaffolding might sustain a new and even better moral order. The science alone won't build it. But we can.

5

Science, Religion, and American Educational Policy

I. Introduction

I take up this topic with some trepidation, for these are often troubled waters. But 'tis oil I mean to spread, and calm I hope to restore. The occasion, of course, is the 1999 decision, by the State School Board for Kansas, to delete the topic of Darwinian evolution from the required curriculum for high school biology, and the topic of big bang cosmology from the required curriculum for high school physics. This dual deletion was itself a gentler event than it might have been, for the worthy group who comprised the school board wisely made no attempt to *prohibit* the teaching of either topic in the high schools of Kansas. Instead, it permanently removed these two topics from the *examinations* that Kansas students are required to write, and pass, to get credit for their courses in biology and physics.

This removal has substantial consequences of its own, to be sure, consequences of which the board was entirely aware. As measured against the narrow goal of getting a decent grade in either of these new "Kansas Science" courses, a student now has no motive (indeed, given finite resources, has a negative motive) to learn the substance of either of the disputed topics, and any teacher who presumes to teach them, in the teeth of the new exam policy, is strictly wasting instructional time that could be better spent elsewhere. This, I believe all parties will calmly agree, was the central *point* of the board's decision. Without imposing an

First presented as the Distinguished Lecture in Philosophy and Public Affairs, SUNY at Albany, April 28, 2000.

outright prohibition on the teaching of either Darwinian evolution or
big bang cosmology, the board nonetheless removed, from both the students and the teachers, *any* short-term personal or instructional motive
for addressing either topic.

My question here is twofold. First, was the Kansas School Board's decision a wise decision, one that other school boards should try to emulate?
And second, if it was *not* a wise decision, *why* was it unwise? What is wrong,
many will ask, with easing a contentious topic or two out of the public
school curriculum, especially through a policy where no actual prohibitions are involved?

If one is a defender of the methods and the achievements of modern
science, one's first impulse here is likely to focus on the comparative *evidence for* or *intellectual merits of* modern evolutionary theory over the available alternatives – the ancient biblical creation stories, most obviously,
or perhaps the so-called creation science currently advanced by some
Christian fundamentalists. That is a natural line of response to take, and
I commend the distinguished people who have pursued it.[1] But it is not
the strategy I shall pursue in this essay, for it unwisely focuses our attention
on the *factual* and *theological* issues at stake instead of where our attention
really needs to be focused, namely, on the *public policy* issues at stake. As a
culturally diverse people, we are likely to be divided by the former; but as
a historical people, we have shown a positive genius for finding an acceptable consensus on the latter. Let us begin, then, by reminding ourselves
of what is truly fortunate about our own past, and what is at least roughly
right about the policies and practices that have governed it.

II. An Argument from Fairness

Thanks first to our founding constitution, America has been free of the
sorts of spiritual and intellectual oppressions often found elsewhere. The
state-driven suppression of all religious activity, as seen in the old Soviet
Union, is something unthinkable in our own country. And so is the state-
driven enforcement of a single religious orthodoxy, the sort of spiritual-
cum-political tyranny that Christianity displayed during the years of the
Roman and Spanish Inquisitions, and that in recent years Islam has displayed in fundamentalist Muslim countries such as Afghanistan and Iran.

---

[1] For example, P. Kitcher, *Abusing Science: The Case against Creationism* (Cambridge, MA: MIT
Press, 1982). Also, R. T. Pennock, *Tower of Babel: The Evidence against the New Creationism*
(Cambridge, MA: MIT Press, 1999).

We occupy a happier position midway between these ugly extremes. Constitutional law guarantees the pursuit of diverse religious practices free from state interference, and it likewise guarantees that no particular religious doctrine or practice, qua religious practice, can ever become established in law as binding on the rest of us.

This much is familiar. But equally important are the details of how this general constitutional policy has played itself out in the domain of educational policy in particular. We, all of us, pay taxes to support a single public school system – there is no escaping this important duty. Even so, America has a long tradition of tolerating – even welcoming – a parallel system of parochial and private schools at all levels, from preschool to research universities. These schools serve the peculiar spiritual and educational interests of any religious group with the resources and the determination to found such an institution. Not infrequently, they provide a better product than their public counterparts. On the whole, they are a proud and prominent feature of our educational landscape. Moreover, and in addition to these "opt-out" institutions, there are the far more numerous weekend or "Sunday" schools. Their function is to provide, as an optional auxiliary to the public curriculum, a more modest measure of sectarian religious instruction for the much larger number of children whose primary educational source will always be, for sheerly financial reasons if for no others, the religiously neutral public school system.

In all, the arrangement works well. Doctrinal enthusiasms of a specifically religious nature do not frustrate and divide us in the design and presentation of our public school curriculum for the simple reason that they are kept out of that public curriculum entirely. The constitutional separation of church and state saves us endless grief and conflict in our grade schools and high schools, just as it does in our social and legal commerce generally. Our inevitably diverse religious practices and enthusiasms are positively protected as a part of each person's inviolably private life. Correlatively, the public institutions, on which we *all* depend, are likewise protected from being co-opted in any respect to serve someone's sectarian and specifically religious purposes. As a nation, we should be moderately proud to have maintained this social equilibrium for so long and so well as we have. And we should view with skepticism any attempt to upset it.

Which returns me to Kansas, and to its school board's recent removal of evolutionary biology and big bang cosmology from the examinable high school curriculum for biology and physics. Was this action intended to serve the religious agenda of America's Catholics, perhaps? Certainly not.

Both topics, especially the former, are widely taught in Catholic parochial schools and universities. Was it intended to protect our Jewish children from heretical ideas? Not for a moment. Almost every sect of Judaism is entirely comfortable with both views. Perhaps the Buddhists, then? Or the Hindus? Of course not. The Hindus have long taught that the universe is extremely old, perhaps even infinitely old. There is no "other-regarding" service taking place here. Quite evidently, the school board's recent action served the doctrinal interests of exactly one religious group: America's fundamentalist Protestant Christians. These are the folks who constitute a majority on the board at issue. And theirs are the religious convictions – namely, an Earth no more than 6,004 years old, and a divine creation, at Earth's inception, for all plant and animal species – which convictions are now spared contradiction by any examination-directed lessons or discussions in the science courses of Kansas high schools. Coincidence? We don't think so.

This is not a hard call. One member of America's religious mosaic has here "broken faith" with the rest of our mosaic, by unilaterally restructuring a part of the *public* school science curriculum in order to serve a patently, specifically, and narrowly sectarian religious purpose. That purpose is the suppression or marginalization of presumptive information that happens to stand in conflict with some of the fundamentalists' own religious convictions. We don't allow other religious groups, large or small, to "sanitize" the public school curriculum according to their own conceptions of cleanliness, and we shouldn't sit still for this ill-advised attempt, by a smallish group of fundamentalist Protestants, to do it either.

I have heard it said, by nonfundamentalists and usually in weary frustration, that the Kansas School Board desperately needs to *take* the very biology and physics courses that they presume to emasculate and, more important still, to take the sequence of college-level and graduate-level courses to which the high school courses lead. This may be true. But this is clearly a partisan retort, and I am here urging a very different point, one that all Americans can embrace. What the Kansas School Board needs is not so much a course in biology or physics. Much more desperately, given its considerable powers, it needs a course in American *civics*. Because, whether by accident or by design, the board has clearly abused those powers.

Is this a hanging offense? Great goodness, no. But it is an offense, and one that wants repairing before the civic felony gets compounded by the passage of time and by the accumulating deficit in the educations of some of Kansas's brightest children. It is also an occasion for reminding ourselves just how our delicate curricular balance is achieved and maintained.

III. Fair Procedures in Curricular Growth

While most of the country has long boasted a public school curriculum that is essentially free of positive religious dogmas, we cannot pretend that there are *no* conflicts, anywhere, between the contents of that curriculum, especially its science curriculum, and the private convictions of some religious group or other. In truth, there are hundreds, even thousands, of conflicts, both large and small. As a nation, we manage to tolerate them without measurable stress. And a moment's reflection will show that we have no choice but to continue to tolerate them. For the attempt to eradicate all such conflicts, by deleting the teaching of any and all topics that encounter a dissenting voice, would reduce a robust, twelve-year curriculum to a withered and empty shell. There is always some parent who is willing to deny the existence of the Nazi death camps of World War II, to insist that the Earth is flat, to deny that microorganisms are the cause of disease, to credit space aliens with the construction of the Pyramids, to deny that the Apollo project landed men on the Moon, or to insist that George Washington never existed. Even mathematics would not be entirely safe. (Apparently, in the early 1900s, one legislator in a southern state proposed a bill to redefine the value of pi as 3.3 exactly, just to tidy things up.) What America has done, in the face of these scattered conflicts, is first to assemble a public curriculum on purely secular grounds, and then to require any dissenters or reformers to ground requests for change in a similarly secular basis. This allows for ongoing change in our school curriculums as human knowledge grows, and it leaves protected the private pursuit of religious convictions. But it also provides the public curriculum with a continuing protection against specifically religious modifications. Those of us who, for religious reasons, wish to make such modifications have to learn to tolerate things as they are unless and until a secular basis can be found to justify them.

Once again, the system has worked well. Our children now move through a curriculum that is much improved and dramatically expanded over the educational substance of generations past. The topics of evolutionary biology and big bang cosmology spring to mind immediately. But they are only a small part of a grand expansion in the sciences generally, an expansion that should be available to all of our schoolchildren, or at least to those who want and need it.

Let us remind ourselves how these topics, among many others, came to be in the Kansas curriculum in the first place, just as they are in the standard science curriculum of every other state in the union and every other first-world nation on the planet. There is no surprise or conspiracy here. They earned their way in, by virtue of having become the consensus core of two of the most successful and vital scientific disciplines of the last century – molecular biology and modern astronomy. They are in the curriculum for the same reason that mathematics is there, and history, and chemistry, and American civics, and literature. They represent some of the most important intellectual achievements that the human race has produced, achievements vital for any student to understand, at least if he or she hopes to have even a passing acquaintance with the current state of humankind's ongoing search for factual knowledge and social understanding.

IV. Fair Procedures in Scientific Growth

Against this observation, the evolutionary and cosmological theories at issue are sometimes portrayed as mere religious enthusiasms themselves, as just more sectarian myths that have no more right to a place in the curriculum than does any other religious claim. If this were true, it would be a serious charge and a weighty reason for confining these theories to the sidelines. But in fact, the charge is quite unfair. Entirely aside from the question of their truth or falsity, the process that led to the consensus place of these theories in the scientific literature (and to their now-dominant guidance of many billions of dollars of research activity, and to their central place in all college-level science curriculums world-wide) is, for better or for worse, deliberately and self-consciously quite *different* from the processes that characterize the history of the world's major religions.

Most important, that consensus-building process reflects once more a vital *separation* between, on the one hand, the private religious beliefs of individual scientists, and, on the other, the collective institutional procedures for *evaluating* the competing theories, experimental data, critical arguments, mathematical proofs, and experimental procedures that make up the ongoing commerce of science. Scientists are people too, and many of them are Buddhists, Jews, Muslims, Catholics, Protestants,

Hindus, Sikhs, and so forth. And they have as much right to their religious, cosmological, metaphysical, and moral convictions as anyone else. But as we academics know well, and as the public needs to be reassured, scientists are expected to leave those private convictions at the door when they – qua practicing *scientists* – enter the collective and deliberately non-private arena of theoretical and experimental evaluation.

This happens, for example, when they write up their theoretical proposals or their experimental results and submit them, in hopes of publication, to one of the prestigious professional journals that form the basic instrument of intellectual communication between all members of the scientific community. Upon receipt of such a hopeful submission, the editor of the journal sends out copies of that written article to several other scientists who are experts in the same research areas as the author. Each of these jurors, or "referees," supplies the editor with a written evaluation of the article in question – of the quality, significance, and likely reliability of the research it presumes to report – and each recommends for or against publication of that article. Most submissions (with top journals, over ninety percent of them) do not make it through this stern filter.

Since the original author has little or no idea to which referees the submitted article will be sent for evaluation (it could be any one of thousands of scholars), he or she must perform the research to be reported, document the procedures used, and frame the scientific arguments put forward, in such a way as to satisfy the most stringent standards of care, knowledge, skill, and argument that the profession can be expected to impose. To attempt to do anything less would be to commit professional suicide. The working life of a typical scientist is mostly taken up by the never-ending business of working to produce and to publish research that is equal to the critical standards of this jurylike "peer-review" process, a process in which each scientist is frequently also a fleeting juror or judge, as well as a regular plaintiff.[2]

Nor does this process of critical evaluation end with publication. Indeed, it then gets tougher. Once the research is thus made available to the entire scientific community, it becomes subject to even broader and deeper criticism. Independent labs set about to test the proposed

---

[2] My professional readers will forgive this brief rehearsal of the rules of the game. They may be familiar to academics, but they are not at all familiar to the general public. The public needs to know that the business of science is regularly conducted in these deliberately nonsectarian and, I hope, objectivizing ways.

theory, or to replicate the published results, with experiments of their own. These experiments may also get published, if they manage to get past the editors described earlier. The fate of one's research (as well as of one's professional reputation) is once again hostage to a process of highly public evaluation. The policy of forcing things into the sunlight is a good procedure here, just as it is elsewhere.

Public – or better, *collective* – evaluation occurs again whenever a scientist presents a research paper to a large professional audience at one of the great many scientific conferences that take place every year. It happens yet again when the scientist applies for a research grant – often in the millions of dollars – to support the laboratory, the instrumentation, and the technical help necessary to pursue a multiyear research program. Here, too, the research proposals must run a gauntlet of multiple peer reviews, and here, also, most applicants suffer disappointment. But the overall result, as the years flow by, is a research industry governed by a family of collective, nonsectarian, and increasingly sophisticated standards of professional evaluation, rejection, and acceptance. Scientists spend almost as much time in preparing to meet the critical evaluations of anonymous colleagues, and in serving as anonymous referees themselves, as they spend in doing research in the first place.

Does this process guarantee that the accepted results of this cognitive industry are certain and factual? Of course not. We should be so lucky. If anything, the most impressive lesson of our scientific history is that major *changes* in our accepted theoretical ideas are the regular outcome of, and a measure of the *health* of, the multilayered institutions of critical evaluation just described. These theoretical or conceptual revolutions seem to happen about once every century for most of the major disciplines. And they are repeated proofs positive that the dogmatic *protection* of its currently dominant theories is not remotely what science is about.

Absolute certainty, then, is at best a distant ideal, no matter how careful our methodology. What that collective critical process does guarantee is only that the ongoing evaluation of alleged research results is as fair, as nonsectarian, and as expertly informed as public human institutions can possibly make it.

While this falls short of providing certainty, it does not fall short of being highly valuable. For one thing, we can all be comfortably assured that, whichever scientific theories may have managed to earn a broad professional consensus, they have not "earned" it by sectarian appeals to anecdotal evidence, divinely inspired texts, uncontrolled experiments, ancestral wisdom, miraculous revelations, pontifical decrees, oracular divinations, dream experiences, answered prayers, selectively mined evidence, or crystal balls. And in particular, even if we are deeply religious ourselves, we can be confident that the current scientific consensus was not earned by appeals to some *other* sectarian religious group's peculiar anecdotes, divine texts, ancestral wisdom, and so forth. We can be confident that the standards that are collectively imposed are standards that look past the idiosyncratic religious beliefs of individual scientists, whatever they might be. The scientific results, therefore, are something in which we can all have some measure of confidence, no matter what our initial convictions. For the methodology that produced them is collectively deployed and religiously neutral, or as neutral as mere mortals can make it.

V. Doctrinal Conflict and Moral Progress

By contrast, of course, the substantive scientific *theories* that emerge from this critical process are often not neutral. Occasionally they stand in flat contradiction to the religious convictions of at least some groups. While the methodology of the sciences may be neutral, its products are often an unwelcome and upsetting surprise to some devout group or other. There is no point in trying to deny, or even to minimize, this important historical fact. The case of Galileo's conflict with the Roman Church over the location and motions of the Earth is perhaps the most famous episode. But there are hundreds, even thousands, of others. What is a wise society to make of such conflicts when they occur, not in our distant and settled past, but in the bustling here and now?

Perhaps not too much. After all, the religions of the world are all in flat logical conflict with one another already on hundreds of doctrinal points – that is why they are distinct religions. But in America, at least, they have learned to tolerate each other, or better, they are *required* to tolerate each other by the various constitutional protections and separations discussed earlier in this essay. If we can already tolerate, without rancor, a substantial diversity of cosmological views, the addition of a handful of new views from the sciences – if and when they are relevant – can only raise the level of intellectual discussion, both within and across faiths.

Raising the level of *that* discussion is also of vital importance, at least for the long run. After all, the great religions of the world have occasionally made substantial changes in their own doctrines and practices, and have occasionally taken major steps forward in their doctrinal wisdom and their moral judgments. Christians no longer burn at the stake the people we used to call witches, even though their Old Testament is still unambiguous in requiring death ("Thou shalt not suffer a witch to live." Exodus, 22:18). The Roman Church no longer burns or imprisons people – as it burned Giordano Bruno and imprisoned Galileo – for teaching that Earth rotates on its axis and revolves about the Sun. Medical vaccinations are no longer prohibited by the church (for a short time, they were) on grounds that they constitute a worldly and profane attempt to interfere with God's will concerning who lives and who dies. Anesthetics are no longer denied to women in childbirth (they were stoutly resisted in some religious quarters) on grounds that they spared women the punishment properly due them for carrying the guilt of Eve's original sin. And, at least in this country, the female population is no longer denied access to higher education, to the voting booth, and to professional careers on grounds that women are God's subsidiary afterthought, made from Adam's rib and destined for lesser duties.

I take it that we can all agree that *these* doctrinal shifts are morally welcome. But we should note that, in every case, the agent of moral change was our growing understanding of and control over the complex natural world. So-called witches (i.e., schizophrenics and manic-depressives) were not "possessed by Satan," they merely had dysfunctional brains, brains we now have some capacity for repairing. Earth is not the "Center of the Universe"; it is a vulnerable speck of dust, cosmologically speaking – one we had best learn to take care of. Vaccinations against disease and anesthetics to curb pain are no more, and no less, violations of God's will than is building a roof to protect ourselves from His rain, or donning clothes to protect ourselves from His cold. And women are not "lesser modifications" of male original material; on the contrary, the Y chromosome characteristic of males (we are XY) is a "piggy-back" device that modifies the basic human XX genetic material, which, unmodified, invariably makes a human female. (If anything, the Old Testament had it exactly backward.) Moreover, the male and female brains produced by human chromosomes are, taken one by one, physically indistinguishable from one another, save for a few tiny nuclei and a handful of chemicals that govern sexual behavior. Certainly they are different in no intellectually or politically relevant dimensions.

The point of reminding ourselves of these historical developments is not to castigate our religions for making mistakes. Science makes mistakes, too. The point is rather that, while religions have a right to protection against persecution, neither they, nor anyone else (science included), has a right to blanket protection against the occasional polite *contradiction*. Even in our public schools. For as we noted earlier, sparing absolutely everyone's children from ever encountering a contrary view on anything would require deleting almost the entire curriculum. And it would prevent our children from slowly entering the ongoing public discussion of new discoveries and theories, a discussion that leads, in the long run, to the precious sorts of enlightened moral and religious *advances* discussed in the preceding paragraph.

VI. An Argument from Practical Consequences

Including uncensored science in the public school curriculum also has an important *middle*-term benefit. Or rather, and to put the point negatively, the sectarian censorship of science instruction, as in Kansas, poses a very real middle-term *danger*. The danger lies in the inevitable incoherence of any Kansas student's comprehension of biological reality, when that subject is instructed in the absence of the only unifying framework that makes it scientifically intelligible and technologically useful. This is dangerous because it will be from this very population of deliberately blinkered students that the next generation of doctors, plant biologists, pharmacists, geneticists, biochemists, agricultural engineers, medical researchers, forensic scientists, and neuroscientists will be drawn.

Collectively, and because of their coherent scientific understanding, these professions do a great deal of good for humanity, in a hundred different dimensions. Correlatively, a significant loss of scientific competence across all these professions would bring to humanity a very real *misery* in all of those same dimensions. For example, the discovery, manufacture, and wise administration of modern medicines is now deeply dependent on our understanding of the DNA molecules that make up the human genome, and in whose chemical structure our evolutionary history, our developmental instructions, and our relations to other species are written. Genetic diseases in the young and the old, the varieties of cancer, the internal activities of our cells, and our endless battle against viruses and bacteria are all problems that require an uncensored grasp of the evolutionary processes that made them, and continue to make them.

We are equally dependent on that knowledge for the wise administration of every aspect of the plant and animal kingdoms, both on the farm and in the wild, as we seek to balance our own needs as a species against the needs of every other form of life on the planet. For better or worse, it has fallen to the human race to be the principal steward of the Earth's biological well-being. *If* the current regime of Nature is a planetary equilibrium achieved and maintained by many interlocking evolutionary processes operating over geological time, then we had best make sure that we understand these processes, and in great detail. We cannot protect what we do not understand. And we will not understand, unless we learn *all* of the relevant science.

Here, incidentally, we should note that big bang cosmology – being our best theory of the genesis of the chemical elements, and of the development of galaxies, stars, and planets throughout the universe – is also a vital part of our planetary understanding. We discard it at our peril. The relevance of general astronomy to the issues at hand is not difficult to see. For example, if the stars were indeed created 6,004 years ago – in 4004 B.C., as biblical literalists reckon the date – and if they began to emit light the instant they were created, then the light from any star in the universe cannot have had time to travel a distance of more than 6,004 light-years from its source. That is to say, no one, anywhere in the universe, would now be able to see anything that lies farther away than 6,004 light-years from their present viewpoint. But most of the stars we see from Earth are reliably measured to be at distances much greater than this. In fact, there is an external galaxy beyond the constellation Andromeda that is now visible even to the *naked eye* on a moonless night, and it is fully one *million* light-years away from Earth. If the creation story sketched earlier were true, then the Andromeda galaxy would not yet be visible, and will not become visible until another 993,996 years have passed. In the absence of modern astronomical knowledge, plainly, one constructs creation stories at one's peril.

Cosmology aside, a deep appreciation and command of genetic theories and technologies is essential again for the wise administration of our own social and legal responsibilities – in criminal investigations (e.g., how to gather DNA evidence), in the courts (e.g., how to evaluate DNA evidence fairly), and in our various corrective institutions (e.g., how to diagnose and repair such things as genetically based defects in emotional profile, social perception, or self-control). A real and uncensored grip on the underlying truths of biological structure, its biochemical embodiment, and its evolutionary history, is essential to all of these undertakings.

In sum, this broad range of complex social duties are not responsibilities to be carelessly entrusted to a new generation of instructionally crippled professionals. But that is the direction in which the Kansas example is taking us. Such an experiment in willful self-destruction – medical, agricultural, social, and legal – is not an experiment that the American people as a whole are remotely willing to perform. On the contrary, a recent poll indicates that the public is quite comfortable with the presence of evolutionary theory in its biology curriculum.[3] More broadly, it is also willing to tolerate the inclusion, in the general curriculum, of courses that explore and discuss the creation stories of a broad spectrum of religions, Protestant fundamentalism included, so long as they are not falsely presented as part of the *science* curriculum.

Kansas, of course, will be more or less safe from the gathering evils of incompetence envisioned in the preceding paragraphs, because the rest of the country, which remains *un*censored in its educational policies, will supply Kansas with the doctors, engineers, and scientists that it needs. And perhaps also, because the proud universities of Kansas will struggle mightily to make up for the instructional deficits and biological ignorance that will now burden its native first-year classes.

But the State School Board of Kansas has nothing to be proud of in any of this. The policy it has so recently imposed on the Kansas school system is unworthy of emulation by anybody. The members of the board have violated a constitutional principle that binds the rest of America together. They have broken faith, not just with the faithless (such as your secularly faithful servant), but with the rest of avowedly religious America. They have needlessly crippled the cognitive development of some of their own students. They have substituted their own sectarian religious judgment for the much more balanced and informed judgment of the religiously diverse community of all the world's scientists. They have begun to endanger the effectiveness of an entire generation of future professionals and public servants, and, thereby, the continuing welfare of the innocent public they will eventually serve. In all, the members of the board have disgraced themselves before a trusting nation. Let us implore them, as our fellow Americans, to reverse their own decision, and free their own children, before the derisive expression "Kansas Science" becomes a new and unwelcome euphemism in the American vocabulary.

---

[3] "Evolution and Creationism in Public Education: An In-depth Reading of Public Opinion" (March 2000), a national survey by DYG, Inc., 36A Padanaram Road, Danbury, CT 06811.

6

What Happens to Reliabilism When It Is Liberated from the Propositional Attitudes?

One of the robustly bright spots in Professor Alvin Goldman's philosophical vision has been his determination to *explain* why some specific cognitive representations should and do count as knowledge, *in terms of* the background reliability of whatever cognitive mechanisms or procedures actually produced those representations on the occasion in question.[1] Beyond giving us some welcome and plausible relief from Gettier-type counterexamples to the original justified-true-belief accounts of knowledge, Goldman's vision here naturally directs our theoretical attention toward the mechanisms that, in living terrestrial creatures, actually give rise to our cognitive representations, and toward the profile of epistemic virtues and vices that those mechanisms may display.

These mechanisms of representation-fixation, and the character of the various representations to which they give rise, are the focus of this essay. But my purpose goes beyond merely plumbing the neural- or implementation-level hardware that serves to execute the molar-level cognitive activities as described by Goldman. In particular, it is not my aim to provide a neural-level account of the fixation of *belief,* for example, or an account of the fixation of any other propositional attitude, for that matter. For what motivates me here is the growing body of evidence that

---

[1] A. Goldman, "Discrimination and Perceptual Knowledge," *Journal of Philosophy,* 73 (1976): 771–91. Also, Goldman, *Epistemology and Cognition* (Cambridge, MA: Harvard University Press, 1986).

This chapter was first presented as a paper at a conference in honor of the work of Professor Alvin Goldman.

the overwhelming preponderance of human knowledge has nothing whatever to do with anything remotely like the propositional attitudes.

Twenty years of research in the several neurosciences, I shall argue later, indicates that the various forms of cognitive representation that dominate, and perhaps even exhaust, human cognition are not the classical propositional attitudes at all. And while they do display important *virtues*, none of these nonpropositional forms of cognitive representation displays the classical virtue of Truth, either – not, at least, as Truth was understood by Tarski. They display interestingly different kinds of virtues, virtues that may serve us better than Truth ever did in providing comprehensive *explanations* of how humans and other animals are so strikingly effective in *navigating* the complex spatial, causal, and social environments in which they live. In sum, I propose to explore how Reliabilism fares, and what form(s) it should take, if we explore a view of knowledge that does not embrace propositional attitudes as its primary representational vehicle, and that does not embrace classical Truth as its primary representational virtue.

I. Motivating an Alternative Vision: Commonsense Arguments

It does not require an enthusiasm for modern cognitive neuroscience to motivate an exploration in the antipropositional directions indicated. Common sense already acknowledges several important varieties of knowledge that are nonpropositional on their face. The many instances of knowledge '*how to*,' provide an important set of opening examples. We learn, and subsequently *know*, how to crawl, how to walk, how to run, how to catch and hit a baseball, how to ride a bicycle or fly an airplane, how to use language and navigate a romantic conversation, how to sing a song or play a musical instrument, how to drive a nail and tighten a bolt, how to write and how to type, how to fight and how to spread oil on troubled waters, and a hundred thousand other skills besides, all of them largely or wholly inarticulable.

To be sure, some parts of the philosophical tradition already celebrate this dimension of human knowledge – Heidegger, Merleau-Ponty, the later Wittgenstein, and Bert Dreyfus come to mind. But we may wish to draw a sharp distinction between mere motor skills and procedural knowledge on the one hand, and strictly factual knowledge on the other, and then honor this distinction by expecting or demanding entirely distinct explanatory accounts of what goes into each of these two types of knowledge. Accordingly, a "justified-true-belief-produced-by-a-reliable-mechanism" account of knowledge need not stand convicted, nor even accused, of explanatory narrowness. *Motor* knowledge can find its own account, no doubt, but from some other quarter.

And yet, such 'monster barring', as Lakatos might call it, depends for its success on the integrity of the distinction it presumes to draw. We must note that the distinction, as just drawn, is still inadequate to the complexity of the situation, for beyond motor skills there is the broad array of *perceptual* skills displayed by humans and other animals. We are able to discriminate and recognize faces, colors, smells, voices, spatial relations, musical compositions, locomotor gaits, the facial expression of emotions, flowing liquids, falling bodies, and a million other things besides, where the exercise of these discriminative skills is once again largely or wholly inarticulable. These skills, too, are learned, often quite slowly. And unlike motor skills, perhaps, perceptual skills are incontestably cognitive skills, and must be an integral part of any account of knowledge.

The classical account, of course, has a well-defined place for them: they are the very mechanisms of singular belief-fixation on whose background reliability our empirical knowledge depends. But this view of our perceptual skills – that they are essentially mechanisms for the fixation of singular propositional attitudes – becomes deeply problematic the instant we direct our attention beyond normal, language-using human adults to the host of other cognitive creatures in the animal kingdom, none of whom use language and all of whom are problematic candidates for the manipulation and fixation of specifically propositional attitudes. An octopus, for example, sees, feels, and hears its way through the kelp and submarine rocks, but it is romantic nonsense to suppose that what its several sensory systems are doing is busily fixing a sequence of discrete propositional attitudes. To put the matter bluntly, the articulation, manipulation, evaluation, and eventual fixation of propositional attitudes is not a game that the octopus has ever learned to play, nor, most likely, ever *could* learn to play. Moreover, the same deflationary estimation is true of every other nonhuman species on the planet, as is illustrated by the extraordinary difficulty encountered in trying to teach any of them the normal use of human *language*, the original and still prototypical system for the expression of propositional attitudes. Whatever the cognitive activity of *non*human animals consists in, it would appear to be something quite other than their deploying an internal, Fodorian-style 'language of thought'.

Very well then, perhaps *only* humans can possess factual knowledge, since only we can fix propositional attitudes...? But this chauvinistic conclusion flies in the face of the manifest perceptual and conceptual sophistication of animals, many of whom have more extensive factual knowledge of their peculiar environmental niches than we outsiders will ever have. And it presumes to draw an indefensible Cartesian-style *distinction* between humans and 'the brutes', indefensible since – language use aside – the brutes display (at least occasionally and at least to some degree) every cognitive capacity displayed by the average primitive human.

What we are struggling with here is the problem of trying to redeploy a particular conception of cognitive activity – the familiar homocentric, linguaformal, propositional-attitude conception – in a variety of domains in which it functions only poorly, or only metaphorically, if it functions at all. This failure might be acceptable, even predictable, if we could mark some fundamental distinction between the respective domains of success and failure, but any such distinction is itself problematic.

It gets more problematic still when we look closely at cognition in the specifically human case. Even here, trying to construe all cases of perceptual knowledge as the fixing of singular propositional attitudes proves to be a real stretch. In the course of a silent hour of industry at my workshop lathe and drill-press – milling, balancing, and then mounting a small leaden disk on the axle of a small, high-speed electric motor to serve as a gyroscopic stabilizer – my perceptual adventures are considerable, and in my engrossed state, they guide my motor behavior at every stage. But I may not have fixed a single propositional attitude in the course of the entire hour. Or, if I did, it was when I allowed myself some brief and tangential thoughts on what time my wife would probably be home for dinner.

The same is true of my perceptual adventures while driving on a long highway trip: I may awaken from a long and fierce reevaluation of Hume's argument for rejecting miracles, only to realize that I can remember nothing of the last twenty miles of perfectly successful highway navigation. If any propositional attitudes were fixed during that period, their topic was Hume's skeptical philosophy, not the details of road and traffic. My perceptual systems were engaged in a very different business: that of directly guiding my ongoing motor behavior.

The same is true of a basketball player when frantically guarding a darting, leaping opponent. The player's visual perceptions of an adversary's every step, hesitation, and acceleration serve to guide his own almost instantaneous motor responses, always tending to block the potential shot, intercept the potential pass, or prevent running access to the basket. The player doesn't have *time* to fix any propositional attitudes, nor time to deploy his deductive talents, so to infer therefrom some appropriate intentions and then act on them, even if he did fix an attitude or two. Anyone who mounts a defense using that cognitive strategy is doomed to a career on the bench. Once again, the human perceptual system is here as active as can be, and active on a typical cognitive task, but the fixing of propositional attitudes seems not to be a part of the process. And yet surely the athlete, like the car driver and the lathe operator, has ongoing perceptual *knowledge* of his physical surroundings. How else explain the aptness of his, and their, motor performances?

Such cases begin to make it robustly clear that, even in the case of *human* cognition, the great bulk of our daily perceptual activity is not plausibly construed as the seriatim fixing of an ever-growing file of discrete propositional attitudes. Such a construal would prompt a number of embarrassing questions in any case. For example, how *many* discrete propositional attitudes (P-As) does our perceptual machinery fix per second? One? Ten? A thousand? Does the rate vary? What is the maximum possible rate of P-A fixation? What happens to them all, once fixed? Further, do our separate sensory modalities simultaneously and independently fix their own sequence of modality-specific P-As? Or is there a central, modality-neutral P-A fixer into which they all feed, either competitively or cooperatively? If the P-A-fixer construal of our perceptual mechanisms is correct, these questions must have determinate factual answers. Providing a plausible set of answers, answers grounded in a detailed understanding of the brain's microstructure, is an undischarged obligation of those who would defend that construal.

I do not wish to claim, at least at this stage, that we *never* fix a P-A in response to our perceptual adventures. But the evidence is clear that such occasional P-A fixings as do take place are ancillary events that are parasitic on a prior or deeper level of cognitive activity, a level that functions continuously and independently of any P-A fixing, a level of cognitive activity that is capable of guiding our motor behavior (indeed, *typically* guides our motor behavior) in the absence of any P-A fixing activities. In sum, our original and still basic mode of perceptual processing has nothing to do with the fixation of propositional attitudes. But it does provide us with knowledge.

We have so far been discussing the fixation of *singular* P-As, but we quickly reach a similar conclusion concerning the role of general or universally quantified P-As within human knowledge at large: their role is secondary, and minimal. One might have assumed them to be central, since collectively those accepted general sentences are supposed to give determinate meaning – at least if one is a Quinean holist about meaning – to the lexicon of concepts that they deploy. But the idea doesn't bear scrutiny. The problem, once again, is the case of nonlinguistic animals – that is, every creature on the planet except us. To judge from the complexity of their perceptual discriminations and motor behavior, they possess conceptual frameworks that rival, even if they do not equal, the conceptual frameworks of humans. And in possessing such conceptual frameworks, they possess, just as we possess, a general knowledge or comprehension of the categories into which Nature divides itself and the structure of the major relations between them. But these animals are *not* plausible candidates for possessing and manipulating a system for the expression of universally quantified propositional attitudes. To put the point bluntly for a second time, the articulation, manipulation, evaluation, and eventual fixation of propositional attitudes is not a game that nonhuman animals have ever learned to play, nor, most likely, ever *could* learn to play. In what, then, does their *general* knowledge consist?

We must ask the same question of ourselves, because when humans suffer the isolated *loss* of their acquired capacity for the expression, manipulation, comprehension, and fixation of propositional attitudes – as happens in *global aphasia* – the bulk of their cognitive capacities remain robustly intact. This stroke-induced neuropathology, long familiar to neurologists, involves massive destruction to the left-side cortical regions surrounding and including Broca's and Wernicke's areas. These areas are vital for the *production* of grammatical speech and for the *comprehension* of grammatical speech, respectively. Their joint destruction leaves a patient who is unable to comprehend speech, either spoken or written, and unable to produce either as well. This deficit is not a superficial perceptual or motor deficit. Such patients can still *sing* snatches of coherent speech, if the song was learned before the stroke. And they can discriminate both voiced phonemes and printed letters as well as you or I. Their deficit is evidently deeper. They have lost their system for expressing, deploying, and manipulating propositional attitudes in the first place. They are out of the propositional-attitude business entirely.

And yet they can still play a game of chess, cook a dinner, appreciate an unfolding football game, drive a car across the state, or shop for the weekend groceries (although the shopping list must be iconic). Such people retain a rich conceptual framework, a rich appreciation of both natural and functional kinds. But not, evidently, because they command

and deploy a Quinean network of accepted general sentences. In what then, does their general knowledge consist?

II. Motivating an Alternative Vision: Arguments from Cognitive Neurobiology

With the millennium finally turned, it is not too soon to assert that we now know perfectly well what the infralinguistic knowledge of humans and other animals consists in, or, more cautiously, what the great bulk of it consists in. Thanks to comparative neuroanatomy and neurophysiology, we know that there is extraordinary conservation of neuronal structures and physiological functions across the brains of terrestrial animals, especially the mammals, and most especially the higher primates. Like it or not, we are all operating with small and mostly continuous variations on the same structural and functional theme. Humans may boast the luxury model among the offerings, but we are all driving four-wheeled internal-combustion vehicles, and the differences between species reflect only local bells and whistles and only modest horsepower gaps.

Cognitive neuromodeling confirms this conformal estimation. When we construct *artificial* neural networks, we discover that these biologically inspired information-processing devices display many of the same avowedly *cognitive* capacities displayed by their diverse biological siblings. These networks are designed to mimic both the transduction sensitivities of our sensory neurons and, most important, the *synaptic connectivity-patterns*, among downstream neuronal populations, shared by all of us. All mammals have retinal and cochlear neurons, for example, and in all mammalian species these sensory neurons project their signal-carrying axons deeper and deeper into the brain, through a series of downstream neuronal populations such as the LGN and the visual cortex, and the MGN and the auditory cortex, respectively. These deeper populations project, in turn, through a series of perhaps a hundred to a thousand further populations, some of which project finally to the body's muscle system, completing what is, at bottom and in all of us, a sophisticated system for sensorimotor coordination.

As anyone who hasn't been living under a rock for the last fifteen years will know, thousands of these variously tuned artificial networks have re-created a broad and coherent family of surprisingly sophisticated cognitive behaviors. These behaviors are achieved not by direct programming, as in the classical approach to artificial intelligence, but rather by a variety of synapse-adjusting *learning* procedures. These procedures gradually shape and reshape the network's pattern of synaptic connections, and thus its cognitive profile, as the network encounters a large series of 'training examples' at its sensory periphery. The networks learn, that is, from their ongoing experience of the peculiar environments they are made to encounter.

Many accessible examples will illustrate both the technique employed and the results achieved. A simple two-layer network, whose input cells model our trichromatic retinal cones, accurately reconstructs the antecedently known dimensionality, organization, and similarity metric of human *phenomenal color space.*[2] A three-layer network crudely modeled on our primary auditory pathway (cochlea to MGN to primary auditory cortex) learned to distinguish *submarine rocks from explosive submarine mines* by processing the sonar echoes returned from each. It achieved a higher level of accuracy than trained human sonar officers.[3] A similar network, crudely modeled on our primary *visual* pathway (retina to LGN to primary visual cortex) learned to recognize, quite reliably, the orientation and several curvature-values of various *multiply curved surfaces,* as presented in sundry gray-scale photographs. That is, it solved the classical 'shape-from-shading' problem in visual psychology.[4] A fourth network, again modeled on our visual pathways, learned to distinguish *female faces from male faces* and to identify specific individual faces across a variety of distinct photographs.[5] Though it had less than 5,000 neurons, its performance-level almost equaled that of humans on the same cognitive task. A fifth network, with a similar architecture, learned to recognize the *eight principal human emotions,* as expressed in a variety of distinct human

[2] L. M. Hurvich, *Color Vision* (Sunderland, MA: Sinauer, 1981); A. Clark, *Sensory Qualities* (Oxford: Oxford University Press, 1993); and P. M. Churchland and P. S. Churchland, "Recent Work on Consciousness: Philosophical, Theoretical, and Empirical," *Seminars in Neurology* 17, no. 2 (1997): 179–86; reprinted in Churchland and Churchland, *On the Contrary* (Cambridge, MA: The MIT Press, 1998), 159–76.

[3] R. P. Gorman, and T. J. Sejnowski, "Learned Classification of Sonar Targets Using a Massively-Parallel Network," *IEEE Transactions: Acoustics, Speech, and Signal Processing* 36 (1988): 1135–40.

[4] S. Lehky, and T. J. Sejnowski, "Computing Shape from Shading with a Neural Network Model," in E. Schwartz, ed., *Computational Neuroscience* (Cambridge, MA: MIT Press, 1988), 452–4.

[5] G. Cottrell, "Extracting Features from Faces Using Compression Networks: Face, Identity, Emotions, and Gender Recognition Using Holons," in D. Touretsky et al., eds., *Connectionist Models: Proceedings of the 1990 Summer School* (San Mateo, CA: Morgan Kaufmann, 1991), 328–37. Concerning the biological reality of the coding strategies here deployed, see A. Hurlbert, "Trading Faces," *Nature Neuroscience* 4, no. 1 (2001): 3–5; and Leopold, D. A., O'Toole, A. J., et al., "Prototype-Referenced Shape Encoding Revealed by High-Level Aftereffects," *Nature Neuroscience* 4, no. 1 (2001): 89–94.

faces.[6] A sixth network, with recurrent axonal pathways to provide a form of short-term memory, learned to distinguish *grammatical from ungrammatical sentences*, as expressed in a simplified but genuinely productive model language.[7] That network accepts sentences of arbitrary length, but just as with humans, the reliability of its grammatical discriminations gradually decreases with the length and complexity of the sentences presented to it.

The point of listing these diverse examples is as follows. In every case, what gets 'fixed' at any point in the network's perceptual processing is *a pattern of activation-levels* across an entire population of neurons. The pattern that gets fixed at the input layer is directly determined, of course, by the nature of the sensory stimulation that hits it. Think, for example, of the pattern of light and darkness projected through your lens onto the surface of your retina, or of the pattern of sound energy at various frequencies distributed along the length of your cochlea.

At the second layer, however, the pattern of activation levels is fixed rather differently. To see this, note in Figure 6.1 that each neuron in the middle layer population of the schematic network receives its own fleeting pattern of excitatory and inhibitory stimulations from the set of axons arriving from the first, or input, layer. These stimulations are variously modulated by the size or 'weight' of the individual synaptic connections through which they have to pass. The resulting activation level of the receiving neuron reflects the 'sum' of those simultaneously modulated incoming stimulations. Every other neuron in that second layer is also the site of a similar 'summation' over its own peculiar family of incoming stimulations. And the overall *pattern* of individual activation levels thus produced across the assembled neurons at the middle layer is the triumphal result of this very first information-processing step. The result is a (new) pattern of activation levels across the neuronal population at the middle layer. We may represent that pattern as a histogram or bar graph, if that is your pleasure. Or as an ordered *n*-tuple or vector of activation values, if you prefer that notation. Or, finally, as a single point in an *n*-dimensional Cartesian coordinate system (with one axis for each neuron in the relevant population), if your intuitions are geometrically

---

[6] G. Cottrell and J. Metcalfe, "EMPATH: Face, Emotion, and Gender Recognition Using Holons," in R. Lippman et al., eds., *Advances in Neural Information Processing Systems*, vol. 3 (San Mateo, CA: Morgan Kaufmann, 1991), 1–7.

[7] J. L. Elman, "Grammatical Structure and Distributed Representations," in S. Davis, ed., *Connectionism: Theory and Practice*, Vancouver Studies in Cognitive Science, vol. 3 (Oxford: Oxford University Press, 1992), 138–94.

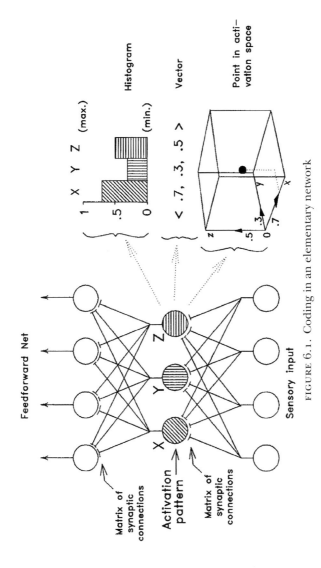

FIGURE 6.1. Coding in an elementary network

97

inclined. These alternatives are severally illustrated on the right-hand side of Figure 6.1. They are just different ways of representing, for our convenience, exactly the same thing: the *activation pattern* across the relevant population.

The story for the third layer is the same, although it will have its own configuration of variously weighted modulating synaptic connections with which to transform the activation pattern it receives from the middle layer, and it will fix its own proprietary activation pattern/vector as a result.

What we are looking at, then, is a multistage device for successively transforming an initial sensory activation vector into a sequence of subsequent activation vectors embodied in a sequence of downstream neuronal populations. Evidently, the basic mode of singular, ephemeral, here-and-now perceptual representation is not the propositional attitude at all; it is the *vectorial* attitude. And the basic mode of information processing is not the *inference* drawn from one propositional attitude to another; it is the synapse-induced *transformation* of one vectorial attitude into another, and into a third, a fourth, and so on, as the initial sensory information ascends the waiting information-processing hierarchy.

That highly trained processing hierarchy embodies the network's *general background knowledge* of the important categories into which Nature divides itself and many of the major relations between them. That is to say, the brain's basic mode of representing the world's enduring structure is not the general or universally quantified propositional attitude at all; it is the hard-earned *configuration of weighted synaptic connections*, those that transform the activation vectors at one neuronal population into the activation vectors at the next. It is these myriad connections that the learning process was originally aimed at configuring, and it is these connections that subsequently do the important computational work in the matured network. We can see how, and in what sense, each family of synaptic connections embodies general information when we look closely at the profound effects it has on the *activation-space* of its proprietary neuronal population.

Cottrell and Metcalfe's face-recognition network (Figure 6.2) provides a prototypical illustration of the phenomenon. Figure 6.3 provides a three-dimensional cartoon representation of the activation space of that network's middle layer (it actually had eighty neurons, not three), at the beginning of training (Figure 6.3a) and at the end of training (Figure 6.3b). Each of the points distributed throughout that space represents the activation-pattern, at the middle layer, produced by one of the roughly

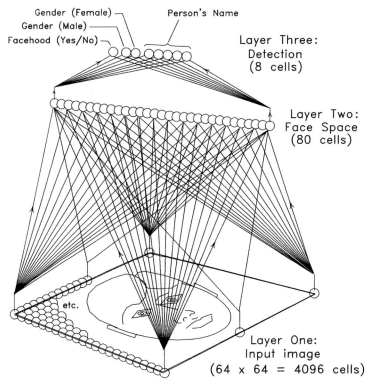

FIGURE 6.2. A network for discriminating faces

100 diverse input images (male faces, female faces, nonfaces), as that sensory input vector was transformed by the matrix of synaptic connections onto the neurons at the middle layer. With the values or weights of those connections set at random, the various input vectors get transformed into a randomly scattered set of middle-layer activation vectors, as seen in Figure 6.3a. There is no organization or acquired wisdom there.

But after training has progressively reset those connection weights to their mature values, the original input vectors now get transformed into an entirely different and highly organized family of middle-layer activation vectors, as seen in Figure 6.3b. Plainly, the network has learned to judge the similarities and differences among its input images in such a way as to discriminate strongly between nonface and face images (the former are all confined to a small subvolume near the origin of the space) and between male faces and female faces (their respective coding regions are mutually closer, but still disjoint), and to group different images of the same person tightly together around a 'prototypical' point for each.

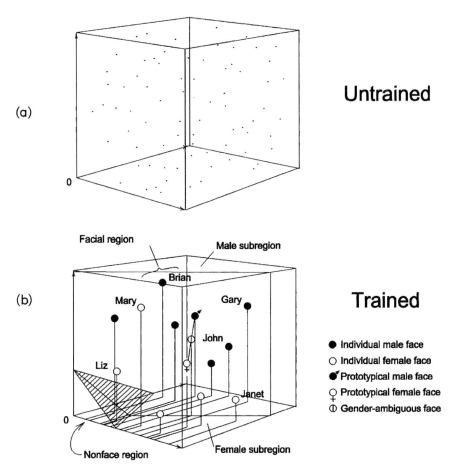

FIGURE 6.3. The genesis of a hierarchical categorial framework

What we are looking at in Figure 6.3*b* is the *conceptual framework* acquired by the face-recognition network. We are looking at its own internal representation of the important classes into which its perceptual environment divides itself, and the most important degrees and dimensions of similarity and difference that variously unite and divide them. We are looking, for what it is worth, at the network's general knowledge of the world's enduring structure.

That conceptual framework, note well, is appropriately hierarchical, in that it displays a subordinate/superordinate structure: for example, *Mary's face* is a *female face* is a *human face*. But it does *not* display the alleged *compositional* hierarchy, so familiar from Locke and Hume, wherein all

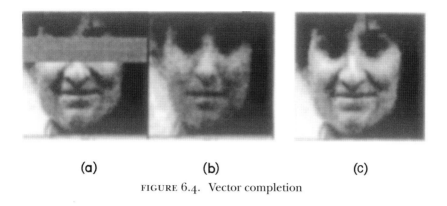

(a)   (b)   (c)

FIGURE 6.4. Vector completion

complex ideas are somehow compounded out of a finite lexicon of conceptual 'simples'. For here there are no 'simples': every concept is a volume – perhaps small if subordinate, perhaps large if superordinate – in a high-dimensional space of distinct similarity estimations. On this view, *all* concepts are complex. Indeed, given the large neuronal populations displayed in biological brains, all concepts are extremely complex.

Note also that this categorical hierarchy embodies a substantial amount of detailed general information concerning the typical or normal instances of each of its categories. This important feature of sculpted activation spaces is hidden by my incidental use of a three-dimensional cartoon to illustrate the organization of learned categories within what is in fact an *eighty*-dimensional space. Just *how much* information has been learned begins to emerge when we present the mature network with a partial or degraded input image, such as the image of Mary with twenty percent of her face occluded by a 'blindfold', as in Figure 6.4*a*. The network will identify her correctly, even so. Indeed, when we examine the middle-layer activation vector produced by that deliberately degraded input, we discover that it contains, though in a highly compressed and encrypted form, exactly the information displayed in Figure 6.4*b*! As you can see, in transforming the input vector into the middle-layer vector, the trained network has produced an activation vector that has 'filled in' the missing sensory information in accordance with its repeated past experiences with undegraded images of Mary. (For comparison, an original training image of Mary is presented in Figure 6.4*c*.) This illustrates the strong and automatic tendency of trained networks to assimilate nonstandard perceptual inputs to the nearest prototypical category available, among those to which it has been trained.

Such 'vector completion' constitutes a kind of abductive inference, as C. S. Peirce would have called it, or an 'inference to the best explanation', as we might be tempted to call it. It shows that during training, the network has somehow come to embody the expectation or information that *all faces have eyes,* and in particular, that *Mary's face always has these specific kinds of eyes.* I use propositional-attitude talk here to emphasize, to our unwashed paleolithic intuitions, the high level of the network's cognitive achievement. But of course that is not how the network represents things. It would be more accurate to say that the network has a robust disposition to react to any partial or degraded sensory input, within hailing distance of some learned prototypical input, as if it were a complete and undegraded instance of the prototype category for which the network already has a complete and high-dimensional representation stored as a learned 'attractor' in the sculpted activation space of its middle layer. Those learned attractors/prototypes clearly contain information about, for example, typical female eyes, noses, mouths, ears, hairlines, jaw shapes, facial proportions, and many other things besides – as attested by the network's filling in any of those details, spontaneously and appropriately, if they happen to be missing from the current sensory input. It has general knowledge of the world's typical or enduring structure, and it automatically deploys that information as a natural by-product of its basic mode of cognitive operation – vector-to-vector transformation.

What the preceding story outlines for us is a systematic conception of human and animal cognition, one grounded squarely in what we know about the neurobiology of mammalian brains and the cognitive behaviors of networks modeled on their structure. It is a conception that offers an alternative to the propositional-attitude conception of cognitive activity embodied in folk psychology and in traditional epistemology. That new conception includes decidedly more than the sketch here provided. It embraces a compelling account of motor skill for example, and of sensorimotor coordination to boot.[8] But the preceding will have to suffice for our purposes. What we want now to address is the issue of how to conceive of *knowledge* in the face of the following awkwardness. None of the presumptive *vehicles* of representation here contemplated – neither *activation vectors* across a neuronal population, nor *synaptic-weight configurations,* nor the resulting *prototype hierarchies* within activation space – are the sorts of vehicles that have a *truth*-value. But the classical account of knowledge

---

[8] P. M. Churchland, *The Engine of Reason, the Seat of the Soul: A Philosophical Journey into the Brain* (Cambridge, MA: MIT Press, 1995), 91–6.

(as justified-true-belief) and its welcome Goldmanian augmentation (as true-belief produced by a mechanism that reliably produces more truths than falsehoods) both require cognitive vehicles that have truth-values. How can we reconcile these competing viewpoints?

III. Representational Virtue without Classical Truth

Reconciliation, of course, *need* not be our aim. We might just toss off truth. Our model here might be that fringe account of classical truth occasionally advanced by the pragmatists, wherein it was proposed to define the truth of any representation directly in terms of the behavioral or navigational successes to which it gives rise: crudely, a true proposition is *one that works*. In the same spirit – but by-passing beliefs, propositions, and truth entirely – we might attempt to define any representation as an instance of genuine *knowledge* just in case, when deployed, it produces successful behavior or navigation.

This tempts me hardly at all. While I am at least a closet pragmatist, and while I resonate to the idea that the brain's cognitive activities are ultimately in the service of motor control, I balk at any such direct definition of the sort proposed. The vagueness of "behavioral success" is one obvious reason. The problem of how to evaluate candidates for knowledge, on this criterion and in isolation from other candidates, when only *collectively* do they play a role in directing behavior, is another. But the most basic reason for rejecting this approach is the broader reason long voiced in objection to the original pragmatist account. Specifically, if we *define* or *identify* what counts as truth, or as knowledge, in terms of the behavioral successes it produces, then we will not be able to give a nontrivial *explanation* of those behavioral successes in terms of cognitive representations that do rise to the level of knowledge and truth. We want to give a *behavior-independent* account of what knowledge is and why it counts as such, at least partly because we want then to be able to *explain* the behavioral success of any given cognitive agent *in terms of* such knowledge that the agent possesses. As well, we want to be able to explain the occasional behavioral *failures* of any cognitive agent in terms of the peculiar defects or absence of knowledge in the agent's cognitive repertoire.

Very well then, I have just talked myself out of a possible solution to my problem. And I have committed myself to the need for a behavior-independent criterion for what counts as knowledge. This might seem to force me back toward the notion of *true* representations, and toward the notion of mechanisms that reliably produce them. But in fact, there are some interesting alternatives here, short of thus giving up on my project.

That there *are* representational alternatives, to propositions sporting truth, is not a matter of any controversy. There are photographs, paintings, architectural drawings, stereo image-pairs, holograms, Fourier transforms, acoustic recordings, logarithmic plots, movie films, algebraic equations (e.g., for straight lines, circles, and surfaces), space-probe image vectors, interference patterns, folding road maps, and a million other things besides, including, note well, sculpted activation spaces and activation vectors at various points within them. What we need to understand here is how activation vectors and activation spaces can *represent* aspects of the world, and what it is for them to do so *successfully*. Only then can we turn with profit to the question of when and how those representational successes are achieved *reliably*.

Let me approach the positive thesis of this essay with a deliberately simplified and easily grasped analogy. We are all familiar with a standard highway map, such as a fold-out portrayal of Los Angeles's freeway system, plus its major nonfreeway street arteries. There is no mystery about how this map successfully represents the city's traffic arteries: there is a unique *mapping*, from the various lines and intersections on the map to the various roads and intersections of the city, *that preserves all of the relative distance relations* between the graphical elements on the map.

No map is perfect, of course, but a road map is accurate exactly to the degree that the relative distance relations that configure the target's road system are faithfully reproduced among the map's graphical elements – its lines, intersections, and so forth. We may not always get perfection in our highway maps, but they are close enough to be useful. Most important, we can say, without mystery or controversy, what successful representation here *consists in* – it consists in the existence of a relation-preserving abstract mapping between the elements of the domain represented and the elements of the domain doing the representing.[9]

Notice that this transparent account of representational success *deploys no causal notions at all*. Mere representational success can strictly be achieved without any causal processes such as those that might be deployed by an army of surveyors or an aerial photographer. My map

[9] Two recent papers usefully press this same theme. G. O'Brien, "Connectionism, Analogicity and Mental Content," *Acta Analytica* 22 (1999): 111–31. Also, G. O'Brien and J. Opie, "Notes Toward a Structuralist Theory of Mental Representation," in H. Clapin et al., eds., *Representations in Mind: New Approaches to Mental Representation* (Westport, CT: Greenwood, 2006),

might have been produced by my accidentally spilling a spoonful of sauce-laden spaghetti on a blank sheet of paper. But if, by chance, the noodles landed exactly like *this* (here I point to an AAA freeway map of Los Angeles), then the image thus created also constitutes an accurate map of the Los Angeles freeway system, accidental though it is. For the relevant abstract mapping is there. Of course, given its accidental origin, you would be unlikely to *trust* the pasta-printed map unless you had somehow verified its isomorphism to a more authoritative map with a more reliable provenance. But I am getting ahead of myself.

A final point about road maps is that one can determine whether two physically distinct maps present the very *same* portrayal of the world, independently of any written labels that those maps may contain, and independently of knowing how, or even whether, those two graphical systems relate to anything in the real world. For one can always determine whether the two maps are isomorphic *to each other*. There is even an effective procedure for making that determination. Make a transparency of each of the two maps so they can be superimposed and effectively compared (Figures 6.5*a* and 6.5*b*). Then take any intersection on map *b* and superimpose it, with a straight pin, on any one of the finite number of intersections on map *a*, as in Figure 6.5*c*. Then slowly rotate map *b*, around the straight pin and over the motionless map *a*, in a search for global congruence of every element in each map. If the two maps are indeed isomorphic with each other, then there must exist exactly one superposition of the map *b* intersection onto some intersection in map *a*, and one rotational position of *b* relative to *a*, which yields a perfect congruence, as is evidently approached in Figure 6.5*d*.

Note that such global congruence gives us a criterion for *map identity* that is utterly independent of any written labels on the map itself, and of any causal relations that either map may bear to any physical road system. Map *b* might have been compiled as a record of a system of *ancient water canals* on the planet Mars, for example, and the evident isomorphism with Los Angeles is an accident. Or both *a* and *b* might have no causal relations to the world at all, being confabulatory maps of entirely fictional cities. But the two maps still embody *identical portrayals* of their target realities, if any. This is an internalist criterion for 'identity of portrayal', or 'sameness of meaning'. It is the nonpropositional analog of what the literature calls "narrow content".

All of this is clear. Now comes the contentious part. If we expand the notion of a 'map' from the familiar case of a structured set of elements on a two-dimensional surface to the more general case of a structured

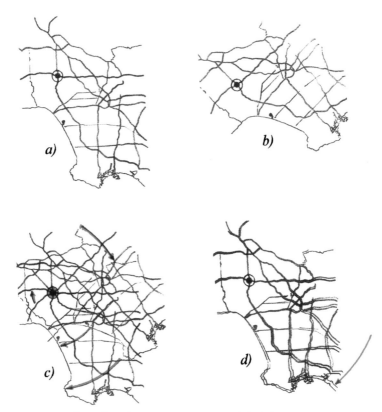

FIGURE 6.5. Rotating two maps to reveal their homomorphism

set of elements in an *n*-dimensional manifold, where *n* can be very large, then we can claim that the various sculpted activation spaces produced in neural networks during learning are all instances of a 'map', in this more general sense of the term. But, in general, they do not map the set of possible positions in *geographical space*, as do the foldout paper maps with which we are most familiar. Rather, those high-dimensional manifolds map the set of possible positions in *human face-space*, as we saw in Figure 6.3, or the set of possible positions in *color space*, or *sonar-echo space*, or *grammatical space*, or *curved-surface space* (recall the list of artificial networks described in Section II), and so on for every other enduring domain of objects and properties that humans encounter, and of which they come to form a conception.

The learned *prototype points* are the principal elements within those high-dimensional maps, and the intricate family of *distance relations* that collectively configure those points is what gives any such map the peculiar (narrow) content or (narrow) meaning that it possesses. For it is these interpoint distance relations that are collectively isomorphic to a family of real-world *similarity* relations among a set of features within the domain being represented. More cautiously, to the degree to which such an iso-morphism does obtain, to that degree is the map a *successful* or *accurate* map of the domain at issue. And, note carefully, it is only the sculpted family of prototypes *considered as a whole* that bears the required isomor-phism to some domain in the world. Individual prototype vectors, taken in isolation, bear no such relations to the world.

Accordingly, what we are embracing here is a new form of meaning holism, one tailored to our new conception of what concepts are. And with it comes a robust criterion for *sameness* of conceptual framework, for *sameness* of world-portrayal: two individuals have the same concep-tual framework just in case the two frameworks embody the same family of distance relations between their many internal elements. That is to say, just in case the two high-dimensional frameworks can be success-fully superimposed to achieve a high-dimensional congruence, as we saw illustrated in the low-dimensional example of the two equivalent freeway maps.

A mature mammalian brain, on this view, is the seat of a large family of sculpted activation-spaces, one for each of the brain's major neuronal populations, and each such space embodies its own intricate family of prototype positions configured by a proprietary set of distance relations. That is what a conceptual framework *is*. And as we saw from the vector-completion properties of Cottrell's face-network, any such sculpted acti-vation space embodies a considerable amount of information about the typical objects or features within the domain that it presumes to repre-sent. That enduring general information is automatically deployed dur-ing perceptual processing (i.e., as the sensory input vector passes through and is modulated by the intervening matrix of carefully tuned synaptic connections) to produce comparatively ephemeral activation patterns across the higher-level neuronal populations. Perception (as opposed to mere peripheral transduction) is thus an inescapably *ampliative* and *theory-laden* activity, even in very simple creatures. For the peripheral informa-tion is always transformed into the creature's waiting categories, and any creature's categorical framework is nothing other than its basic *theory* of how the world is put together, just as Quine taught us. What Quine missed was the nonpropositional nature of our most fundamental theories. But good pragmatist, naturalist, and reductionist that he was, he might have found this wrinkle welcome enough.

Quine aside, how will *Goldman* greet this (admittedly large) 'wrinkle'? I don't know. Goldman is a harder sell, at least where these wares are concerned. But hope springs eternal, and I shall bring this essay to a close by arguing that, at least where his familiar Reliabilism is concerned, he can welcome this new epistemological framework without significant change. Indeed, what I really want to suggest is that the epistemological framework here outlined can *subsume* his *belief*-focused Reliabilism within a more general but still welcoming framework.

When does a sculpted activation space – that is, an entire conceptual framework – count as (general) knowledge of the world? As Goldman's position suggests, it must meet two conditions. First, it must bear the relevant relation-preserving mapping to the relevant domain in the world. That is to say, it must be representationally successful. And second, it must not be a sheer cosmic accident that it does so. The process or mechanisms that produced that successful representation must be a process that reliably constructs a roughly accurate portrayal of the world.[10] Knowledge implies some degree of authority, and authority, whether direct or inherited, presupposes some background reliability somewhere. Here we must appeal to the integrity of those chemical, biological, and physiological processes that normally shape the configuration of our 10^{14} synaptic connections. 'Appealing' to their integrity is about all we can do at this point, since we understand them so poorly. But at least we have escaped our profession's monomaniacal preoccupation with rules for the acceptance or rejection of propositional attitudes. What we need to study instead are synapse-adjusting processes such as Hebbian learning, and, beyond these, we need to study the general theory of dimension-reduction and information-compressing procedures.[11] From these alone will we understand how a creature parlays its ongoing sensory activity into a general theory of the world. For the classical, P-A-fixation approach to learning obviously presupposes a system of concepts *already in place*. Explaining

---

[10] For reasons similar to Goldman's, I here let go the traditional requirement that the relevant representation must be 'justified'. That requirement is often awkward to impose, and what little work it does is better done anyway by the 'reliable mechanism' condition here contemplated.

[11] A recent issue of *Science* (290, no. 5500 [Dec. 22, 2000]) contains three very useful papers on these topics. For an introduction, see H. S. Seung and D. D. Lee, "Cognition: The Manifold Ways of Perception," 2268–9. For more detail, see J. B. Tenenbaum, V. de Silva, and J. C. Langford, "A Global Geometric Framework for Nonlinear Dimensionality Reduction," 2319–23. Also, S. T. Roweis and L. K. Saul, "Nonlinear Dimensionality Reduction by Locally Linear Embedding," 2323–26. These outline two different procedures for extracting simple structure hidden in complex data.

the origins or development of such a system is therefore forever beyond the reach of that approach.

Let us now turn our attention to singular activations *within* a creature's encompassing conceptual framework. In particular, when does an ephemeral activation vector, produced in perception here and now, count as knowledge? It is reasonable to require, as with general knowledge, that it be representationally successful. But here we have an interesting problem. It emerges when we contrast two important classes of activation vectors: those produced by sheer energy transduction at the absolute sensory periphery (at the rods and cones of the retina, for example, or the hair cells of the cochlea), and those produced by subsequent, synapse-induced *transformations* at successive neuronal populations downstream from the sensory periphery.

The former or epithelial vectors are unproblematic. An activation vector at the retina represents just as a photographic image represents, and similarly for a cochlear vector, although that will be a one-dimensional instead of a two-dimensional 'image'. In these cases, as with the other epithelial vectors, there is a relation-preserving mapping from the elements of the representing vector to the elements of the reality represented. Here, of course, the mapped elements are ephemeral rather than enduring.

The downstream vectors, however, do not wear their representational intentions so boldly on their sleeves. They are the progressively transformed offspring of those transparent initial vectors. And during the passage through one, two, or fifteen intervening matrices of synaptic connections, the information in these original vectors has been convolved, compressed, and encrypted into a form that is intelligible only to the rest of the idiosyncratic network of which it is a part. There is thus no obvious way of identifying any relation-preserving mapping that holds directly between the internal structure of these higher-level activation vectors and the reality that they represent. Precisely because they are so wonderfully and fruitfully 'encrypted', their representational contents have been made opaque, at least to one's superficial gaze and when considered in isolation.

But then let us *not* consider them in isolation. Let us ask, *into what* general scheme of things have they become encrypted? And the answer is, into that coding scheme embodied in the larger space of *possible* activation-vectors, of which they are just so many fleeting instances. More to the point, they have been effectively encrypted into that encompassing space which embodies a carefully sculpted family of *prototype regions*, that space which embodies an entire *conceptual framework*, that space whose internal elements and their characteristic proximity relations *do* map intelligibly into some domain of objects, features, and similarity relations in the world.

Our earlier analogy will make this clear. Like a single point chosen at random on a foldout highway map, there is little or nothing about the intrinsic properties of that point that bestows upon it the geographical significance that it clearly has. Nor do its raw coordinates on that piece of paper (e.g., "G-17") have anything directly to do with it (in printing, the road system might just as well have been rotated 45 degrees relative to that rectangular grid). What gives that point its peculiar representational significance is the family of *distance relations* it bears to all of the *other* elements of the map – its sundry lines, curves, areas, and intersections. With regard to the representational significance of road-map elements, one is plainly *obliged* to be a semantic holist.

And so, for similar reasons, is one obliged to be a semantic holist where concepts (read: prototype points or regions) are concerned. They are the salient points within our high-dimensional activational manifolds, and like the salient elements of any map, they derive their meaning from their *collective* portrait of the general and enduring features of reality. When we do have genuine knowledge of the ephemeral perceptual world, it is only because we see it through the lens of our acquired conceptual frameworks. Plato would appreciate this result, and so might Kant, despite its antinativist basis.

There is another irony here. Most of our profession, from Locke and Hume to the present, have assumed that meaning flows *from* the singular and the sensory *to* the general and the nonsensory. Careful and resourceful theorists such as Dretske and Fodor have spent most of their careers trying to capture semantic meaning in terms of the 'indicator' features of ephemeral perceptual tokens. On their view, concepts (token-*types*) get their meaning from the causal provenance of their ephemeral tokens, and each token-type enjoys its peculiar meaning independently of whatever other concepts may or may not lie in the surrounding conceptual neighborhood. For them, that is, meaning is acquired atomistically rather than holistically.

Though well motivated, perhaps, this view is plausible only if you stick closely to concepts applicable in spontaneous perception. Even then, it fails to account for the idiosyncratic ways in which different individuals may happen to *conceive of* the object of their knowledge. And finally, the view could never begin to account satisfactorily for *chronic* errors in our conception or perception of the world, since our chronic responses are supposed to be precisely what bestows a specific meaning in the first place.

We can now see why it suffers these failures. In fact, semantic content or meaning flows in precisely the opposite direction. It flows *from* the general and the enduring *to* the singular and the ephemeral. What my activation vectors 'mean' is fixed by the comparatively enduring elements and structures within the sculpted activation space that is their inescapable theater of operations. The assembled structures within that very space collectively embody the peculiar conceptual 'take' on the world that learning has produced in the relevant creature. A singular error of perceptual judgment occurs when the current configuration-of-objective-features that caused a current activation vector is not the configuration-of-objective-features assigned to that point in activation space by the background mapping that grounds the conceptual framework as a whole. And chronic errors of perception typically occur when the background framework itself is faulty in some way. Our perceptual machinery is then doomed to encrypt its epithelial sensory vectors into a portrait or map of the world that positively *mis*represents its enduring objects and features. We speak in such cases of chronic misconceptions (e.g., an immobile Earth) and chronic misperceptions based upon them (e.g., that the setting Sun moves downward).

With these things sorted out, we can finally readdress the question of when a nonepithelial activation vector counts as knowledge. First, it must enjoy representational success – as we have newly come to understand it. That is to say, the singular objective situation that occasioned that vector in the here and now must be an instance of the general type of objective situation assigned to it by the abstract background mapping that grounds the conceptual framework as a whole. Second, it must have been produced by a mechanism of vector-fixation that is generally *reliable* in producing activation vectors that are successful in the sense just outlined.

Here we need *not* wave our hands in ignorance. Those mechanisms of activation-fixation are precisely the matrices of synaptic connections that learning has so carefully shaped. We know roughly how they work, and we can model their activities in revealing detail. It is already plain both how and why they *are* extraordinarily reliable – both in their immunity to scattered neuronal dysfunction (because their representations are highly distributed and their processing is massively parallel), and in their ability to surmount noise and signal occlusion from the outside world (because, once trained, they bring to the business of perception substantial prior knowledge of how the world is put together).

My conclusion, then, is that terrestrial creatures, humans included, quite regularly have knowledge. For the representational success condition and the reliability condition are both quite regularly met, at least for many of the domains we encounter. But I close with a caution. On this view, we are also likely to be the victims of chronic *failures* of knowledge, including (inevitably) chronic failures of *perceptual* knowledge. The reason is straightforward, and it repeats a theme familiar to us from Paul Feyerabend's writings,[12] and from the very earliest of my own writings.[13] On the view proposed in this essay, one's perceptual representations are only as good as the peculiar background conceptual framework in which, for each individual, those representations are destined to be expressed. And such a framework is always and ever a fragile, an imperfect, and, in the long run, an *ephemeral* attempt to grasp the structure of the world. Plato's timeless, ideal heaven is, in the end, a poor metaphor for one's general conceptual framework. A sunlit morning's hopeful spiderweb would be a more accurate metaphor.

[12] P. K. Feyerabend, "Explanation, Reduction, and Empiricism," in H. Feigl and G. Maxwell, eds., Minnesota Studies in the Philosophy of Science, vol. 3 (Minneapolis: University of Minnesota Press, 1962). Also, Feyerabend, "How to Be a Good Empiricist – A Plea for Tolerance in Matters Epistemological," in B. Baumrin, ed., *Philosophy of Science: The Delaware Seminar*, vol. 2 (New York: Interscience Publications, 1963), reprinted in B. Brody, ed., *Readings in the Philosophy of Science* (Englewood Cliffs, NJ: Prentice Hall, 1970).
[13] P. M. Churchland, *Scientific Realism and the Plasticity of Mind* (Cambridge: Cambridge University Press, 1979), chap. 2.

7

On the Nature of Intelligence

Turing, Church, von Neumann, and the Brain

Abstract: Alan Turing is the consensus patron saint of the classical research program in AI, and his behavioral test for the possession of conscious intelligence has
become his principal legacy in the mind of the academic public. Both takes are
mistakes. That test is a dialectical throwaway line even for Turing himself, a ter-
tiary gesture aimed at softening the intellectual resistance to a research program
which, in his hands, possessed real substance, both mathematical and theoretical.
The wrangling over his celebrated test has deflected attention away from those
more substantial achievements, and away from the enduring obligation to construct a substantive theory of what conscious intelligence really *is*, as opposed to
an epistemological account of how to tell when you are confronting an instance
of it. This essay explores Turing's substantive research program on the former
topic, and argues that the classical AI program is not its best expression, nor even
the expression intended by Turing. It then attempts to put the famous Test into
its proper, and much reduced, perspective.

I. The Classical Approach: Its Historical Background

Alan Turing wanted to know, as we all want to know, what conscious intelligence *is*. An obvious place to start one's inquiry is the presumptive and
prototypical instance of the target phenomenon – normal humans – and
the endlessly clever and appropriate behaviors they display in response
to the endlessly various perceptual circumstances they encounter. Conscious intelligence presents itself, from the outset, as being some sort
of *enduring capacity*, possessed by humans and at least some other animals, to generate behavioral outputs that are somehow appropriate, given
the prior state of the intelligent system and the sensory input it has just
received.

This modest opening observation, to which all may agree, leaves us confronting two substantial problems. First, how do we specify, in a suitably general way, what the relation of *input–output appropriateness* consists in, a relation that has infinitely many potential instances? That is to say, what is the peculiar behavioral or functional profile that is our explanatory target here? And second, what sort of internal capacity, power, or mechanism within us is actually responsible for *generating* those appropriate outputs, when given the sensory inputs in question? These are both very hard questions. Given the complex and open-ended character of the input–output relation at issue, we are unable, at least at the outset, to do much more than lamely point to it, as "the one displayed by normal humans." Indeed, it may turn out that the only way to provide an illuminating answer to our first question is to specify in detail the internal generating power or mechanism whose nature is queried in our second question. For a general appreciation of how that mechanism works, and of how it actually generates its remarkable outputs from its sundry inputs (including, occasionally, some null inputs), will automatically give us a finite but general specification of the infinite appropriateness-relation at issue.

Now, to a mathematician such as Alan Turing, and at a time of mathematical development such as the middle of the twentieth century, this situation has some obvious and intriguing parallels with a large class of similar situations in the domain of computable functions. For a simple example of the relevant parallel, consider the mathematical function familiar to us as "the basic parabola," namely, $y = x^2$. This dictates, for any given "input" value for x, a unique "output" value for y. Indeed, this function can be usefully and accurately seen as an infinite set of ordered pairs, $< x, y \ (= x^2) >$, as found in the partial listing that follows.

$x,$	$y \ (= x^2)$
$\vdots$	
$< -2,$	$4 >$
$< -1,$	$1 >$
$< 0,$	$0 >$
$< 1,$	$1 >$
$< 2,$	$4 >$
$< 3,$	$9 >$
$< 4,$	$16 >$
$< 5,$	$25 >$

< 6,   36 >
< 7,   49 >
⋮
etc.

One cannot specify this function by writing out the entire list of the ordered pairs that it embraces, for the list would be infinitely long. Indeed, given that the function includes the real numbers as well as the rationals, it is nondenumerably infinitely long. How, then, do we get a grip on it? Fortunately, we can specify, in finite compass, a *recursive procedure*[1] for *generating* the right-hand member of any pair given the left-hand member as input.[2] Any schoolchild can execute the relevant procedure, for it involves nothing more than addressing the relevant input number and then multiplying it by itself to yield the relevant output. The recursive procedures involved in multiplication may take some time to execute, if the numbers involved happen to be very large. But the time taken to complete the procedure is always finite, for the numbers involved are always finite. While the set of ordered pairs indicated earlier is indeed an infinite set, every left-hand number in every ordered pair it contains is nonetheless a finite number, and thus the right-hand element can always be generated, as indicated, in finite time.

Accordingly, the specification of a recursive procedure for generating, or computing, the output appropriate to any input, given that input, constitutes a *finite* specification of the entire infinite set at issue. It gives us a finite but still-firm grip on an infinite, but sufficiently ordered or well-behaved, abstract reality.

Except for these finite, recursive specifications, any infinite set, such as the entire set of < *x, y* > pairs for the basic parabola, would be forever confined, beyond our cognitive reach, to Plato's nonphysical Heaven.

Since any function whatever is a set of ordered pairs, we can now divide the class of functions into two mutually exclusive subclasses – those that

---

[1] Intuitively, this is any rote procedure for transforming input items into further items, a procedure that contains subprocedures that can be deployed again and again until some criterion is met, at which point the procedure then halts with the production of an output item. Note, to *recurse* is literally to *go back and do again*. Prototypical examples of recursive procedures are the ones you learned for the addition, subtraction, multiplication, and division of largish numbers.

[2] Strictly, if the input is an irrational number, we can only *approximate* the output, but we can do that arbitrarily closely. I will ignore this qualification henceforth.

admit of a finite but possibly recursive specification, in the manner of our parabolic example, and those that do not. The former are called *computable* functions, and the latter are called *noncomputable* functions. These latter possess insufficient order, structure, rhyme, or reason to admit of any specification more compact than a literally infinite list of their ill-behaved elements. They are, in short, utterly unspecifiable by anyone short of God himself.

Turing's basic theoretical suggestion here is that the general input–output relation that characterizes normal human behavior-in-the-world is one instance of a computable function. After all, our behavior-in-the-world displays a *systematic* if complex structure, and the brain is quite evidently a *finite* system. The guess that human conscious intelligence is, in some way or other, a finite computational specification of an infinite set of potential input–output pairs, is at least an intriguing entry point for further research. Specifically, this guess sends us in search of the very computational procedures that the brain presumably employs in generating its behavioral magic.

There is no hope, of course, that the relevant procedures will be as simple as those deployed to compute the elements of the function $y = x^2$. None. But that is not what drew us, nor Turing, in this computational direction. What drew us was the need to find some finite way of specifying the infinitely populated appropriateness-relation discussed earlier, and the need to find a substantive explanation of how its outputs are actually generated from its inputs. The computational suggestion owed to Turing promises to solve both problems at once. If we pursue a research program aimed at recovering the computational procedures actually used by the avowedly physical brain, or procedures functionally equivalent to them, we can hope to provide a finite but nonetheless general specification of the infinite Platonic set that presumably constitutes human rationality, a set that exists, in its entirety, only in Plato's Timeless Heaven. At the same time, we can hope to provide a nuts-and-bolts account of how (some of) the elements of that infinite set are (occasionally) computed in this, our physical/temporal world. This, as I see it, is Turing's basic and most enduring insight on this issue.

It is not, to be sure, Turing's only insight. Students of the history here will point immediately to Turing's characterization of what is now called a universal Turing machine, a physically realizable discrete-state device that he showed to be capable, at least in principle, of implementing any recursive procedure whatsoever. Given Alonzo Church's prior characterization of computable functions as exactly those functions whose input–output pairs are finitely specifiable by some recursive procedure, Turing's demonstration concerning the recursive prowess of his universal machine entailed that such a machine was capable, at least in principle, of computing *any computable function whatsoever*. This is often referred to as the Church-Turing thesis.

This is indeed of extraordinary importance, both for good and for ill. It lies at the heart of the development of the programmable digital computer, and it lay at the heart of the first great wave of research into what quickly came to be called artificial intelligence. It is not hard to see why. What we now call a "computer program" is just a sequence of instructions that directs the (universal) computer's hard-wired central processor to execute this, that, or some other recursive procedure, so as to generate appropriate outputs from whatever inputs it may receive. A finite machine is thus rendered capable, depending on what recursive procedures we choose to program into it (a comparatively simple matter), of generating the elements of any computable function we might wish. That is to say, one and the same (universal) machine is capable of imitating – or better, capable of temporarily becoming – any "special-purpose" machine we might wish: a spreadsheet calculator, a word processor, a chess player, a flight simulator, a solar-system simulator, and so forth. As we all know, that is how things have turned out, and it is entirely wonderful.

More to the point, the Church-Turing thesis also entails that a universal computer – which, plus or minus a finite memory, is what any standard desktop machine amounts to – is also capable, at least in principle, of computing the elements of whatever marvelous function it is that characterizes the input-output profile of a *conscious intelligence*. Hence the rationale for the original or "classical" approach to creating an artificial intelligence: *find/write the program that, when run on a universal computer, will re-create the same input–output profile that characterizes a normal human*. Given only the modest assumption that the input–output function characteristic of conscious intelligence is *not* one of those pathological, unspecifiable, noncomputable functions, the Church-Turing thesis appears to guarantee that the classical research program will succeed. For it guarantees that there exists a finite specification, in the form of a recursive procedure, that will generate any element of the Holy Grail here at issue, namely, the infinitely membered function characteristic of a conscious intelligence. Our only task is to find that procedure, if only by successive approximation, for it is guaranteed to be there.

II. The Classical Approach: Its Actual Performance

All of this is strictly correct, and the relevant program of research, now forty years old, has produced a great many marvelous things. But, curiously, a programmed machine that displays conscious intelligence, or anything close to it, is not among them. Researchers began, understandably enough, by attempting to re-create only this or that isolated *aspect* of conscious intelligence, such as the ability to construct logical or algebraic proofs, the ability to segregate a perceptual scene into discrete objects, the ability to parse a sentence, the ability to navigate a toy environment, and so forth. This divide-and-conquer strategy was entirely reasonable: let us first figure out how the obvious components of intelligence are, or might be, achieved, and then worry about integrating our partial successes later on.

Early successes were plentiful, and encouraging. A good machine, and a clever program, can produce dazzling novelties. But as the decades unfolded, the attempts to achieve progressively greater faithfulness to the actual perceptual, cognitive, and behavioral capacities of humans and other animals proved to require an almost exponential increase in the processing times and the memory capacities expended. Progress slowed dramatically, and attempts at suitable integration were postponed indefinitely. All of this was darkly curious, of course, because the speed of signal conduction inside an electronic computer is roughly a million times faster than the speed of signal conduction along the filamentary axon of a human or an animal neuron. And the evident "clock speeds" of neural computing elements were also a million times behind their electronic brothers. The advantage should lie entirely with the electronic machine.

But it didn't. And whatever was going on inside the human brain, it became increasingly implausible to picture it as engaged in the same sort of *furiously* recursive activities known to be heating up the memory registers and central processors of the electronic machines. The Church-Turing thesis notwithstanding, a reluctant but real skepticism began to spread concerning the classical approach described. How could biological brains be achieving so much more than the high-speed computers, when they apparently had so much less to work with? And why was the classical program hitting a brick wall?

The answer has largely to do with the decidedly *special-purpose* computational architecture of biological brains, as we are about to see. But that is only part of the answer. In retrospect, the classical or program-writing research program presents itself as embarrassingly presumptuous in one central respect. It assumed that the Platonic Function characteristic of conscious intelligence was more or less available or accessible to us *at the outset.* It assumed that we already knew, at least roughly or in general outline, what the membership of the relevant infinite set of input–output pairs really is. It remained only to reprise or re-create that evident profile by means of suitably recursive procedures. So put the programmers to work.

But that Platonic Function (let us agree with Church and Turing that such a computable function exists, if only in Plato's Heaven) is not in the least evident and available to us. It is only dimly grasped by the folk notions that make up our commonsense "folk psychology," and those notions address only a small fraction of the full range of cognitive phenomena – that is, the full extent of the infinite Platonic set – in any case. To assume, as the classical approach did assume, that the target function is more-or-less known by us, and is at least roughly captured by commonsense folk psychology, is to wrongly privilege at the outset a narrow, partial, and possibly false conception of our target phenomenon. It also turns our attention away from the various kinds of empirical research that might positively help in finally *identifying* our target function – research, for example, into the computational activities of the visual system, the motor system, the somatosensory system, and the language system. To deliberately turn one's back, as the classical research program did, on *computational neurobiology* is to engage in the most egregious form of self-blinkering. For whatever the target function might be, and however else it might be finitely specified, we know going in, thanks to Alan Turing, that it is *already finitely specified in every adult human brain.* For the adult human brain is already happily engaged, somehow or other, in computing it. If a finite specification of our target function is what we want to get our hands on, the human brain is the uniquely authoritative place to find it!

Unfortunately, there was a widely repeated and highly influential argument *against* this presumption of unique authority. In retrospect, it can be seen to be a very bad one. Looking at the intricacies of the biological brain, ran the argument, is like looking at the intricacies of a computer's hardware: it won't tell you where the *real* action lies, which is in the peculiar program that the hardware happens to be running.

This breathtaking argument makes the unwarranted and question-begging factual assumption that the biological brain is *also* a general-purpose or universal machine, relevantly like a standard digital computer, a machine that acquires a specific cognitive profile only when programmed with specific, and comparatively ephemeral, recursive procedures.

But the brain is no such thing. And it has been known to be no such thing since the very beginning of the classical research program, when John von Neumann, who is primarily responsible for the architecture of today's serial-processing digital computers, published his seminal researches into the similarities and the differences that unite and divide standard electronic computers and natural biological brains. Published back in 1956 and entitled *The Computer and the Brain*, that prescient little book concludes that the presumptive computing elements within the brain, namely, the neurons and their synaptic connections, are both *too slow* in their activities, and *too inaccurate* in their representations, ever to sustain the sorts of high-speed and hyperrecursive activities required for the success of a discrete-state serial-processing electronic computer. The brain is not remotely fast enough to perform, one after another, the thousands upon thousands of recursive steps that are as natural as breathing for a standard computer. And the imperfect accuracy with which the brain would perform each such step would inevitably lead to fatally accumulated and magnified errors of computation, even if it were fast enough to complete them all in real time. In sum, if the brain were indeed a general-purpose digital serial computer, it would be doomed to be both a computational tortoise and a computational dunce.

As von Neumann correctly perceived, the brain's computational regime appears to compensate for its inescapable incapacity for fast and accurate logical "depth" by exploiting instead an extraordinary logical "breadth." In his own words, "... large and efficient natural automata are likely to be highly *parallel*, while large and efficient artificial automata will tend to be less so, and rather to be *serial*" (emphasis his). The former "will tend to pick up as many logical (or informational) items as possible *simultaneously*, and process them *simultaneously*" (emphasis mine).[3] This means, von Neumann adds, that we must reach out, beyond the neurons, to include all of the brain's 10^{14} *synapses* in our all-up count of the brain's "basic active organs." This means, allow *me* to add, that the biological brain can execute a walloping 10^{14} elementary computational operations all at once: not in sequence – all at once. And it can do so ten times a second; perhaps even a hundred times a second.

---

[3] J. von Neumann, *The Computer and the Brain*, new ed. (New Haven, CT: Yale University Press, 2000), 51.

Von Neumann's observations here were completely correct. But he might as well have been whistling in the wind. As was quickly pointed out, the Church-Turing thesis still guarantees that, however the idiosyncratic human brain may manage its own biologically constrained internal affairs, the function it embodies can nevertheless be computed by a suitable program running on a universal computer. And for aspiring assistant professors in computer science, writing trial programs on a modern machine was much easier than doing empirical research into the brain's microanatomy and computational physiology. Hacking away at skill-specific programs became the Normal Form for research into artificial intelligence. Von Neumann's book, effectively written on his deathbed, faded into obscurity, and was not reprinted for almost half a century. Most important of all, a great gulf was rationalized, and then fixed, between the empirical brain sciences and the discipline of "AI." Cross-fertilization fell close to zero. Neither discipline, it was agreed, had anything vital to teach the other. Despite that disconnection neuroscience went on to flourish. But by the late seventies, AI had begun to stagnate.

III. Turing's Real Legacy

Can we blame any of this on Turing himself? I think not. Despite being co-responsible for the Church-Turing thesis, Turing does not marshal the methodological arguments, scouted earlier, that counsel turning our backs on the biological brain as a fertile route into *identifying*, and subsequently re-creating a recursive procedure for *generating*, the infinite function for conscious intelligence. Nor does he pretend that we already know, at least well enough, what that function is. On the contrary, Turing closes his famous paper "Computing Machinery and Intelligence,"[4] written in defense of the possibility of machine intelligence, by recommending that we simply *give up* the ambition of writing recursive procedures for adult human intelligence directly. He suggests, instead, that we approach the goal of such procedures in the same way that the biological brain does: via a long period in which it gradually *learns* the set of adult skills and capacities we seek to re-create. On this approach, a successful machine intelligence will acquire its sophisticated skills and capacities not in one fell swoop, from the program-downloading activity of some implausibly omniscient hacker. Rather, it will learn those capacities from its continual interactions with the world in which it has to live, just as brains do.

---

[4] *Mind* 59 (1950) : 433–60.

Guided by Turing, our aim has now shifted from writing the input–output program for an adult human, to writing the program for an *infant* human. The Church-Turing thesis provides the same guarantee, presumably, that a universal machine will once again be capable of computing the relevant computable function. That program, as Turing remarks, will have to be capable of some form of self-modification or self-modulation if it is to be equal to the task of mastering the vast curriculum that confronts it. (This introduces some nice wrinkles because, like a leopard, a function strictly cannot change its spots: it is the same infinite set of ordered pairs at one time that it is at any other time. But that is no problem in principle. The input and output elements of any of its ordered pairs must now contain subelements, some of which index the current but modifiable state of the computing system. This leads us into the domain of what are called *dynamical systems*.)

Well and good, but on this importantly revised research program, something is dead obvious that was not so obvious on the classical research program that dominated AI from the nineteen sixties to the nineteen eighties. I complained a few pages back that classical researchers assumed, quite wrongly, that their target function was more or less *known* to them, at least in its important outlines. The justice or injustice of that complaint aside, no one will pretend that the target function for a newborn human *infant* is even remotely known, especially when that function includes one of the most poorly understood capacities in the entire human arsenal – the general capacity for learning itself. How can we hope to get a grip on the elements of *that* infinite function, as a clear target for our attempts at recursive reconstruction, short of going directly to the empirical well of brain development, neuronal and synaptic plasticity, and general cognitive neurobiology? As urged earlier for the adult brain and the adult function, the infant brain is the uniquely authoritative source from which we might learn or recover the infant function. For we know, going in, that the infant brain must embody a finite specification that allows it to compute the elements of that function. For compute them it does, right out of the box.

Turing's closing advice thus leads us back, immediately, to the very same empirical coal face that was deliberately forsaken by his presumptive intellectual heirs. I wish to suggest, therefore, that Turing's usual depiction, as the patron saint of the *classical* research program in AI, is simply a mistake. He is more accurately seen as the unsung patron saint of the more recent and biologically inspired program of research into *artificial neural networks*. This alternative but now flourishing approach attempts to find out both the *what* of the brain's abstract functional endeavors and the *how* of their actual physical computation.

These artificial models portray the brain as a massively parallel *vector processor*, as a nonserial, nondigital computer that transforms high-dimensional input vectors (namely, the pattern of activation-levels across a large population of neurons) into output vectors (patterns of activation across a downstream population of neurons), which ultimately control the body's muscle system. The vectors are transformed by the peculiar configuration of synaptic connections that both separate, and join, one neuronal population to another. For those vectors get appropriately transformed into new vectors, when they are forced to traverse the matrix of synaptic connections at issue. Collectively, those synaptic connections embody everything the creature has ever learned, a gradual process that involves the successive modification of the connection strengths at issue, modifications made in response to the creature's ongoing interactions with the environment.

This alternative picture is as computational as you please. It is another instance of Turing's basic theoretical insight into our capacity for conscious intelligence. Specifically, a well-trained brain embodies a finite specification of a potentially infinite range of input–output pairs – that is, a computable function – a finite specification that takes the form of computational procedures for the repeated transformation of inputs into outputs. This picture also addresses squarely the fundamental issue of how the brain *learns* (a matter mostly finessed by the classical tradition), just as Turing's closing discussion advises. The difference between the two traditions lies mainly in the kinds of representations involved, and the kinds of computational procedures deployed. But Turing would have welcomed high-dimensional vector/matrix processing as eagerly as any other computational device. I therefore suggest that the true heirs to Turing's basic theoretical suggestion are those who pursue the research tradition of artificial neural networks and its fertile interaction with the empirical neurosciences. For that is where Turing's unambiguous advice, tendered as the conclusion to his most famous paper, now bids anyone go.

IV. Reevaluating Turing's Test

What is the status, from this reworked perspective, of Turing's (in)famous behavioral test? (I shall here assume the reader's familiarity with it.) Certainly the computer's interactive behavior is *relevant* to the question of its conscious intelligence, in the way that an arbitrarily chosen finite subset of a given infinite set can at least occasionally *falsify* their ascription to some target function. On the other hand, being finite, the set of input–output pairs revealed during the Turing test always *underdetermines* the claim that they belong to the target function, though they may provide some degree of corroboration. The claim of a successful reconstruction of conscious intelligence is therefore always and ever subject to future refutation. This is entirely normal for *any* hypothesis, and Turing was entirely aware of it. The point behind *Turing's* sketch of the test situation was surely to bring home to his readers that, if they choose to withhold the ascription of intelligence to a computing machine that "passes" his test, they are prima facie denying the efficacy of the very same sorts of evidence that license that same ascription to normal humans. Dialectically speaking, this puts the ball in the doubter's court, who is thus invited to explain and justify this disparity in evidential treatment.

Readers will recall that, at that point in the article, Turing turns to canvass a long sequence of precisely such exemptive apologias, each of which he finds inadequate to blunt the initial presumption that sufficiently systematic intelligent behavior, despite its nonstandard computational source, still has its normal evidential relevance. As a dialectical strategy, this is entirely understandable, and it requires us to ascribe neither more nor less authority to his famous test than we would ascribe to any other, inevitably fallible, inductive/abductive inference.

Is this as close as we can ever get to authoritatively identifying genuine instances of conscious intelligence? No. We can get closer. But to do so, we need to gain an understanding of what naturally occurring conscious intelligence *is,* an understanding that runs much deeper than we currently possess. In particular, we need to know what computational procedures the brain is actually deploying, so that we may have (1) a better grip on whatever infinite function it may be computing, and (2) a better understanding of how the output elements of that function are physically generated, within us, from its input elements. With such a deeper understanding safely in place, we can then address any novel candidate for the possession of conscious intelligence by examining its *internal computational procedures*, in order to get a more authoritative judgment on the identity of whatever abstract function it may be computing, and a more authoritative judgment on whether it deploys the same transformational *tactics*, as deployed in the human case, in order to compute that function.

These deeper probings, note well, will still leave us with an importantly ambiguous situation, a residual problem that lies behind – far behind – the value or legitimacy of the Turing test. Specifically, in the ascription or denial of conscious intelligence to any novel physical system, which criterion should dominate: sameness of abstract function computed? Or (more stringently) sameness of computational procedures actually deployed in the generation of that function's elements? To my knowledge, Turing never came down firmly on *either* side of this question, despite the orthodox expectation that he would opt for the former position. For my part, I am inclined to embrace the latter position. This is not because I wish to exclude nonstandard critters from the ranks of the blessed. My concerns, indeed, are inclusive rather than exclusive. The fact is, no two of us normal humans are computing exactly the same abstract function. Its existence, as that which unifies us, is a myth. Individual variation in our cognitive profiles is ubiquitous, even worthy of celebration. But we *do* share relevantly identical arsenals of computational machinery: crudely, vector-coding systems and matrix-multiplication systems. What unites us all, in the end, is our sharing the same basic kinds of computational machinery. That empirical machinery, and the endless forms of articulation it may find in various individuals and in various species, is the true subject of the cognitive sciences. If we seek the essence of our endlessly variable Natural Kind, that is surely where it lies – not in Plato's Heaven, but inside the head.

8

Neurosemantics

On the Mapping of Minds and the Portrayal of Worlds

I. Introduction: The Problem

A perennial problem in the philosophy of language, and in the theory of mind, concerns the proper criterion for mapping the lexicon of one language onto the lexicon of another, or the concepts of one person's conceptual framework onto the concepts of another's, in such a fashion as to preserve sense, meaning, or semantic identity across the pairings effected by such a mapping (see Figure 8.1). This "translational" problem is part and parcel, of course, of the larger ontological problem of what meaning is and of what concepts are, and thus it is unlikely to be solved independently of some correlative account of both of these background matters. Disagreements on the former topic are sure to be entangled with disagreements on the latter topics, and so it is with those of us who defend a *state-space semantics* (SSS) approach to these problems against those who champion a *language-of-thought* (LOT) approach. For SSS theorists, concepts are functionally salient *points, regions,* or *trajectories* in various neuronal activation spaces; for LOT theorists, concepts are functionally salient *wordlike* elements in a *languagelike* system of internal representations. For both groups, however, the plausibility of their favored approach depends, in part, on the integrity and plausibility of the inevitably quite different accounts of "translation" that they provide.

The present paper takes up these issues as they are variously developed in three recent papers.[1] My first aim is to defend a criterion for

[1] J. A. Fodor, and E. Lepore, "All at Sea in Semantic Space: Churchland on Meaning Similarity," *Journal of Philosophy* 96, no. 8 (1999): 381–403; E. Tiffany, "Comments and

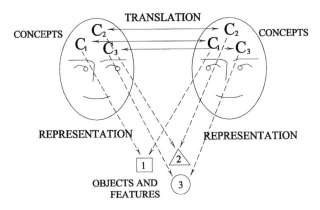

FIGURE 8.1. A schematic of the problem. What is the relation that maps identical conceptual frameworks across individuals? And in virtue of what relation(s) does a conceptual framework portray the objects and/or features of the objective world?

identity of meaning, and a measure of *similarity of meaning*, across distinct conceptual frameworks, a criterion that makes no appeal to any causal or informational connections that the concepts at issue, or their idiosyncratic neuronal dimensions, may bear to properties in the surrounding environment. On this occasion then (in contrast to Churchland 1998), I will articulate and defend a strictly *internalist* account of sameness-of-meaning, an account that leans only on the geometrical similarities between the internal semantic structures of the two conceptual frameworks being compared. I will also address the consequences of this internalist tilt for the proper aims of a semantic theory. As these aims emerge in what follows, they are quite different from the classical aims pursued by Fodor and Lepore. Specifically, I wish to challenge both the conventional wisdom and Fodor and Lepore's wisdom on how it is that a conceptual framework *portrays* or *represents* the world (see the dashed lines in Figure 8.1). I mean to defend an alternative approach to this question, an alternative that denies the notion of reference, and the machinery of model-theoretic semantics, the central role that tradition assigns them.

Criticism: Semantics San Diego Style," *Journal of Philosophy* 96, no. 8, (1999): 416–29; P. M. Churchland, "Conceptual Similarity across Neural and Sensory Diversity: The Fodor/Lepore Challenge Answered," *Journal of Philosophy* 95, no. 1 (1998): 5–32.

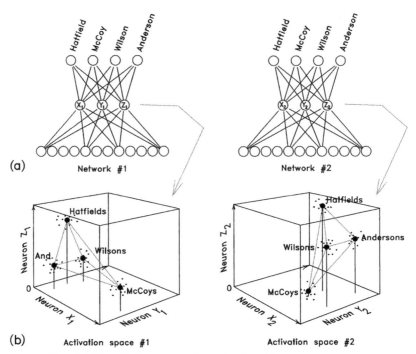

FIGURE 8.2. (*a*) Two distinct networks trained to discriminate photos of faces as belonging to one of four hillbilly families. (*b*) The two activation spaces of the respective middle layers of the two networks. Each has acquired a structured family of "prototypical" family regions, within which facial inputs from each of the four families typically produce an activation pattern.

II. Reliable Translation without Appeal to External Causal Connections

I continue to maintain that one can measure the degree of similarity between two conceptual frameworks, as argued earlier,[2] and as illustrated in Figure 8.2. Despite expectations, this can be done without regard to the causal/informational/semantic significance of any of the axes of either of the two neuronal activation spaces involved – the two spaces that embody, respectively, the two frameworks being compared. In particular, it does not require that either of these two spaces must share any axes with the *same* significance. Indeed, typically they will share no such common axes.

Admittedly, however, the several so-called prototype points (see Figure 8.2*b*) have here been "labeled" in accordance with the real-world

---

[2] Churchland, "Conceptual Similarity."

faces – the Hatfields, the McCoys, and so forth – that *causally produce* the sundry activation-patterns in the neighborhood of those preferred points. And it was these co-labeled alien points that were initially *paired* so that the relevant families of distance relations within each point-family could then be measured and compared across the two networks. Without some such initial pairing of the appropriate prototype points, we will not be able to measure and compare the corresponding *distances* between the several corresponding points. Our similarity measure will not be able to get off the ground.

The same point must be conceded for the similarity measurements made across the Laakso-Cottrell color-discriminating networks,[3] which served as the principal examples of my 1998 paper. The various point-families being compared across those two dozen networks had each point firmly identified by the input stimulus that gave rise to it. And points were paired across distinct networks by virtue of their being *caused* by exactly the *same* external color. And it was only because the points in one family could thus be uniquely paired, with the points in the other family, that Laakso and Cottrell could then identify and measure the corresponding distances that collectively configure each point-family within its propri-etary space.

As Evan Tiffany understandably complains, however, helping oneself to such *direct* causal/semantical identifications of the prototype points looks like cheating.[4] After all, the account at issue has been offered as an internalist account of meaning-similarity. (And I have already forsaken any *in*direct appeal to the causal/semantical significance of any of the *axes* of the activation spaces being compared.) But in fact – and this is the principal claim of this section – such direct causal/semantical identifica-tions of the prototype points are wholly unnecessary to the business of mapping the conceptual structure of one network onto the conceptual structure of another, and equally unnecessary to measuring the degree of similarity that, failing perfect identity, each structure bears to the other.

Let me illustrate with the example already in use – the two imaginary face-recognition networks that have been trained to discriminate mem-bership in one of the four hillbilly families. Let us banish the family-name labels from the four prototype points within the activation space of each network and confront those two irregular tetrahedrons naked of any

---

[3] G. Cottrell and A. Laakso, "Qualia and Cluster Analysis: Assessing Representational Sim-ilarity between Neural Systems," *Philosophical Psychology* 13, no.1: 77–95.
[4] Tiffany, "Comments and Criticism," 426, first paragraph.

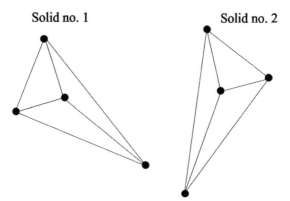

FIGURE 8.3. The two configurations of facial prototype points, one for each of the two trained networks. Note the absence of semantic labels on any of the points.

causal or semantic identification whatsoever, as portrayed in Figure 8.3. How shall we identify and map the appropriately corresponding vertexes (the "prototype" points) across these two irregular solids? Without labels to guide our pairings, we might seem to be stuck. But in fact, the very opposite is the case. By hypothesis, these two irregular solids are metrically *identical*. Their respective internal distance relations are the same. And that fact allows us to exploit the following procedure.

Take the longest edge of each of the two solids and superimpose them as in Figure 8.4*a*, dragging the rest of each solid along behind. There are two possible orientations here, one rotated 180 degrees relative to the other. Of these, choose the orientation (Figure 8.4*a*) that finds pairs of *coplanar* alien points (or the largest number of such pairs), where the planes are *normal* to the superimposed longest edges. Now simply *rotate* one of the two solids around their common longest-line axis, as

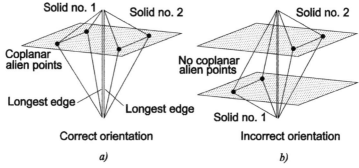

FIGURE 8.4. Two possible alignments of the two solids at issue, alignments that superimpose the longest edge of each.

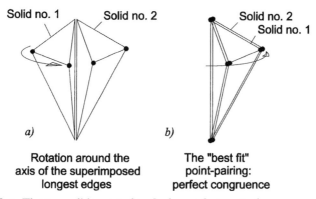

Rotation around the
axis of the superimposed
longest edges

The "best fit"
point-pairing:
perfect congruence

FIGURE 8.5. The two solids rotated to find a perfect mutual congruence of corresponding vertexes and edges.

in Figure 8.5*a*, until the several distances between those coplanar alien points fall simultaneously to zero, as in Figure 8.5*b*. At this unique relative orientation, the two solids will be perfectly congruent. Each vertex and each edge will find a uniquely corresponding alien vertex and alien edge. Since the two solids are metrically identical, there must be at least one such congruence-producing mutual orientation. And since the solids are irregular (i.e., neither solid is individually self-symmetric under any rotation), there must be at most one such mutual orientation. Figure 8.6

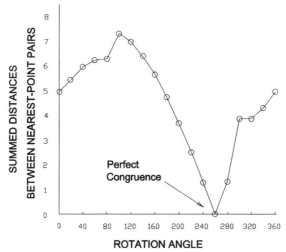

FIGURE 8.6. A graph of the summed distance-from-perfect-congruence, for each of the nearest coplanar vertexes, as a function of the rotation angle, around the longest edge, between the two solids.

graphs the distance-from-congruence (strictly, it graphs the changing *sum* of all of the distances between nearest coplanar alien pairs) against the rotation angle for the two irregular tetrahedrons at issue. Note the unique orientation that achieves a distance measure of zero.

Accordingly, and at the end of this congruence-finding procedure, the prototype points that get paired by sheer spatial superposition are the ones that are *semantically identical*, according to the account of meaning to be defended in this paper, of which more anon. For now, note that the discovery or specification of such *same-sense pairings* of prototype points depends in no way on any antecedent semantic labelings rooted in causal connections to the external world. None whatsoever. We are here contemplating a purely internalist account of semantic identity. This is, very roughly, the SSS analog of what the literature calls "narrow content."

You now have the basic picture to be exploited in what follows, but let me clean up some details before moving on. First, the instruction to superimpose the longest side of each solid (prior to sweeping for coplanar pairs and then rotating to find congruence) will in some cases encounter the difficulty that each solid has *two* or more "longest" sides of *equal* length. This poses no problem. Simply execute the preceding procedure for each of the possible superpositions. On the suppositions of the preceding paragraphs, there is only one orientation of perfect congruence to be found, wherever one might begin the search.

Second, since the mirror image of any hypersolid also counts as an equivalent conceptual framework (because it has the same family of internal distance relations), the search procedure described earlier must be performed, in every case, with the *mirror image* of one of the two frameworks *also* being fitted against the other. This is because, famously, mirror-image figures, such as left and right hands, cannot be brought into mutual congruence unless one of them first undergoes a mirror inversion.

Third, the mapping procedure at issue will be effective for hypersolids or hyper-point-families of arbitrarily large dimensionality. Nor will it matter, to repeat a point from my 1998 paper, if the hypersolids being compared come from still higher-dimensional embedding spaces of different dimensionalities. Indeed, this will typically be the case.

Fourth, the initial restriction to *irregular* hypersolids (those that have no self-symmetry under any rotation) is neither particularly presumptuous factually nor very important semantically. In fact, one has to work very hard to train a network of any size to achieve a perfectly symmetrical arrangement among its prototype points. With nonlinear transfer functions embodied in every neuron, and millions of diversely valued synaptic connections, self-symmetric hypersolids are perfectly possible, but spectacularly unlikely.

Withal, self-symmetric hypersolids remain possible, if improbable. What is their significance? Not much. Two networks, each harboring the same self-symmetric hypersolid within its activation space, would present a rare case of "translational indeterminacy," a case where there exist two equally good mappings of the one framework onto the other, two distinct but equally faithful – indeed, equally *perfect* – interpretations of their respective "narrow contents." A simple representational analogy for this unusual possibility would be a pencil drawing of an ugly bearded face with a hyperbolic hairdo, a drawing that presents qualitatively the same bearded face and hairdo when turned upside down because what was originally the "hairdo" becomes, when inverted, the "beard," and vice versa. The existence of such self-symmetric pictures does not undermine the general notion of "same picture." Neither should the parallel possibility trouble our notion of "same conceptual framework."

So far we have been discussing the case where the two prototype-point families being mapped are metrically identical. But this case is an ideal, rarely realized in reality. How, then, can we hope to pair up "same-sense points" in real cases? Exactly as before, but with lowered expectations. The appropriate procedure for real cases is the same procedure used for the ideal cases, except that now the outcomes fall short of finding a *perfect* congruence between the two hypersolids at issue, and the pairings effected signal only similarity of meanings, rather than identity.

The hunt for the "best fit" (short of perfect congruence) may require one to explore more than one edge-pairing prior to mutual rotation in search of a best fit. But the best fit will still be distinguished by its producing the lowest possible "summed distances from perfect congruence" measure, as illustrated earlier in Figure 8.6. That measure will not be zero, but it may still represent by far the best possible fit between the two superimposed hypersolids. And once that best fit is determined, the "corresponding" vertexes and edges of the respective hypersolids can be paired. One can then apply directly the zero-to-one similarity measure of my 1998 paper, or the GPA similarity measure originally used by Laakso and Cottrell.

Finally, should there happen to be, for some pairs of imperfectly matching hypersolids, two or more *equally* low minima of this sort, then we will confront once again a case of translational indeterminacy, a case of two or more equally good (or, more likely, equally *bad*) translations of one representational medium into another. Such cases lead by stages to the quite real possibility that there is *no* mapping from one framework to the other that will lead us to anything but translational frustration, no mapping that produces a similarity measure of better than, say, one-half. Such pairs of frameworks approximate what the literature calls "mutually incommensurable" conceptual frameworks.

The upshot of this section is that we can perfectly well define and use a purely internalist notion of sameness and similarity of *configurations-in-activation-space* for prototype-families across distinct neural networks. Further, we can define what it is for two prototype points in distinct networks to occupy metrically identical or metrically proximate positions within their respective prototype-families. And we can do all of this *without* knowledge or consideration of any causal connections that they *or* their underlying neuronal axes may bear to aspects of the external world. If Tiffany, or Fodor and Lepore, had any residual worries on this score, they can safely put them aside. The remaining question, which F&L urge as their principal objection to my 1998 paper, is whether this machinery has anything to do with *real concepts* and *real meaning*. The substance of their complaint is that the account of meaning-similarity proposed in my 1998 paper is an *ignoratio elenchi*. To this matter, I now turn.

III. The Conflict Put in Focus

As a basis for meaning, a family of distance relations within a hyper-space may seem altogether too austere to be plausible, especially when contemplated in isolation from any causal connections to the external world. And, most certainly, this picture does contrast with F&L's popular alternative picture of a set of mutually independent and semantically unstructured atoms – the basic concepts of the "language of thought" – where each atom is firmly tied to some aspect of the external world by some proprietary causal link, without which link the atom would be without any content at all.

I mean to celebrate this contrast, not to minimize it. There is at stake here an issue, or a tension, that has been with us at least since Frege. It centers on the contrast between meaning as reference, extension, or denotation versus meaning as sense, intension, or connotation. For F&L, it is meaning as *reference* or *denotation* that is basic and primary. For F&L, the *content* of a concept is simply the real-world object or feature to which, thanks to evolution, it is causally connected. The concept-holder's *beliefs*, should he or she happen to have any, are strictly irrelevant to meaning.

Moreover, for F&L, internal structure for concepts is neither necessary nor desirable. And for them, it is a minor mystery why anyone ever *mis*understands anyone else, since all normal humans share a common conceptual lexicon that is causally connected in all of the same ways to all of the same features in the world. And each such person, note well, is thus canonically connected to the world without ever lifting a cognitive finger to try to understand it or to represent its objective structure. For F&L, a meaningful conceptual framework is something that comes for free to each of us, something whose explanation lies, perhaps, in the evolutionary history of the species.

For me, by contrast, meaning as *sense* is the primary phenomenon. For me, the *content* of a concept is its highly peculiar *portrayal* of some aspect of the world, a portrayal that is often quite inaccurate, a portrayal that enjoys no automatic referential connection to the external world simply by virtue of having a semantic content in the first place. For me, the details of a creature's beliefs or "cognitive portrayals" of the world are downright constitutive of meaning. For me, even a so-called basic concept has a rich internal structure, a structure that is entirely necessary to the peculiar bit of world-portrayal that it embodies. For me, it is a minor marvel that anyone ever understands anyone else, since we all have slightly different world-portrayals, and with them, slightly different connections to the world. But each of us, note well, has made those connections the old-fashioned way – we *earned* them by strenuous cognitive activity expended over years of learning, a process whose principal product is a fairly stable portrayal of the major categories into which Nature divides itself and the chronic relations that unite them into a single, structured world. For me, this meaningful conceptual framework – this world portrayal, this *theory* – is a real epistemological achievement, something whose explanation lies not in the history of the human genome, but in the peculiar cognitive history of each individual, and in the peculiar cognitive history of the society in which that individual was raised.

Because these doctrinal differences have divided us for a quarter of a century now,[5] since long before neurocomputational models began

---

[5] For the major elements of the exchange, see J. A. Fodor, *The Language of Thought* (New York: Crowell, 1975); P. M. Churchland, *Scientific Realism and the Plasticity of Mind* (Cambridge: Cambridge University Press, 1979); J. A. Fodor, "Observation Reconsidered," *Philosophy of Science* 51 (1984); P. M. Churchland, "Perceptual Plasticity and Theoretical Neutrality: A Reply to Jerry Fodor," *Philosophy of Science*, 55 (1988); J. A. Fodor, "A Reply to Churchland's 'Perceptual Plasticity and Theoretical Neutrality,'" *Philosophy of Science*, 55 (1988); J. A. Fodor and E. Lepore, "Paul Churchland and State-Space Semantics,"

to intrude themselves into epistemology and semantic theory, it is no surprise that F&L and I view that intrusion with rather different eyes. F&L ask, reasonably enough, what structured neuronal-activation spaces might have to do with concepts and meaning, as conceived by them and other semantic atomists. And they suspect, again reasonably enough, that the answer is, "Not very much." But that is because they are holding up the wrong explanatory target to begin with. As I shall try to show, structured neuronal-activation spaces have *everything* to do with concepts and meaning, as conceived by semantic *holists*.

Here, incidentally, I disagree with Tiffany's claim[6] that the SSS approach is entirely *neutral* as between causal/correlational, teleological, and holistic approaches to the issue of semantic content. The SSS approach is not just a story about what counts, at the physical level, as a semantic *vehicle*. On the contrary, the explanatory resources held out to us by neural-network models of cognition bid fair to provide us with a systematic intertheoretic reduction/explanation of *holistic* semantic theories in particular. This reduction relocates our current, linguaformal and rather superficial, understanding of cognitive and semantic phenomena within a much broader explanatory framework. That framework reaches out to include the cognition of prelinguistic infants and nonhuman animals, and it reaches down to make contact with the microanatomy and physiological activities of the brain. A relevant dialectical analogy, then – which we connectionists hope to live up to – is the historical conflict between the Ptolemaic and the Copernican/Keplerian accounts of planetary motion. With the subsequent development of Newtonian dynamics – that is, Newton's general theory of motion and gravitation – the heliocentric account received a vindicating intertheoretic reduction, while the geocentric account was unmasked as a fairy tale. Here, on the semantic front, it is the vectorial kinematics of active neurons and the vector-processing dynamics of neural-network models that promise a similar explanation and vindication of semantic holism over F&L's semantic atomism. Let us now examine how it might do so.

---

chap. 7 of *Holism: A Shopper's Guide* (Oxford: Blackwell, 1992), reprinted in R. N. McCauley, ed., *The Churchlands and Their Critics* (Oxford: Blackwell); P. M. Churchland, "Fodor and Lepore: State-Space Semantics and Meaning Holism," in McCauley, *The Churchlands,* 272–7; J. A. Fodor and E. Lepore, "Reply to Churchland," in McCauley, *The Churchlands,* 159–62; P. M. Churchland, "Second Reply to Fodor and Lepore," in McCauley, *The Churchlands,* 278–83; Churchland, "Conceptual Similarity," 5–32; and Fodor and Lepore, "All at Sea," 381–403.

[6] Tiffany, "Comments and Criticism," 417.

IV. Neuronal Space versus Semantic Space

Fodor and Lepore object that, even if my 1998 similarity measure[7] should provide a well-behaved measure of similarity between the *brain* states of distinct individuals, there is no good reason to regard it as providing any measure at all of similarity between their respective *semantic* states. I grant that any such assimilation must be earned, not just claimed. But F&L misconstrue the assimilation being claimed. Whatever else it might be, the measure at issue (hereafter, SIM) is most emphatically *not* a measure of similarity between global *brain* states, as F&L seem determined to see it.

The whole point of that measure (and of Laakso and Cottrell's cognate GPA measure) is to address a highly abstract and overtly functional aspect of distinct individuals. It is an aspect that two individuals might share despite having wildly *different* brain states, different in every one of the trillions of internal synaptic connections that respectively structure them, different in all of their vectorial responses to the same environment, and different (perhaps by billions) even in the numbers of neurons that they possess. In fact, the cognitive systems being compared need not be brains at all. They need only meet the abstract conditions imposed on being a connectionist network. Specifically, they must represent with fleeting vectors of some kind, and they must process those vectors by some realization of the abstract business of multiplying those fleeting vectors by comparatively enduring or stable matrices.

Human vectors are (realized in) patterns of activation across large populations of neurons, and human matrices are (realized in) vast configurations of variously weighted synaptic connections. But any number of alternative physical realizations are possible, especially in the electronic realm. And SIM is a measure of conceptual similarity that will address all such realizations, directly and with indifference to their physical idiosyncrasies. If a cognitive state has to be *abstract* and *functional* to count as a semantic state, then a structured family of prototype points within an activation space is hardly disqualified. On the contrary, it will be among the first theoretical candidates worthy of our consideration.

---

7 The idea here is extremely simple. For any two paired edges, AB and A′B′ (one from each of the two solids being compared), divide the difference in length between them by the sum of the two lengths. This will always produce a fraction between 0 and 1, tending toward 0 as the two lengths are identical, and tending toward 1 as the two lengths diverge. Now take the average of all such fractions, as computed across all of the paired edges, and subtract it from 1. Canonically, similarity (SIM) = 1 − avg.[| AB − A′B′ | / (AB + A′B′)].

The abstract nature of connectionist accounts of cognition is not a new point. (It receives extensive discussion in McCauley's 1996 collection.[8]) But the point merits repeating in the face of connectionism's regular misportrayal as a bottom-dwelling *implementation*-level account of cognition. In fact, the real implementation-level account here is not hard to find: it is empirical neuroscience, the study of the microanatomy and the physiology of terrestrial brains. A salient *virtue* of connectionism, however, over other abstract, molar-level cognitive accounts, is that it makes systematic explanatory *contact* with that microanatomy and physiology. Specifically, large neuronal populations can implement high-dimensional vectors, large populations of synaptic connections can implement the vector-transforming matrices, and the gradual modification of those synaptic weights can implement the process of learning. In addition, so far it is the *only* molar-level account to make any remotely plausible explanatory contact with the implementation-level account of modern neuroscience. That is partly why many of us find it so intriguing.

Still, a version of F&L's question remains: why see a structured family of so-called prototype points in an abstract activation-space as reconstructing anything to do with concepts and their meanings? The reasons are many, and systematic, and they are compelling primarily in concert. Let me now pursue the positive story, with occasional asides concerning F&L's competing account.

V. Structured Activation Spaces as Conceptual Frameworks

The first thing to say about prototype-families is that they are not innate but learned, learned from the network's repeated encounters with instances in its perceptual environment. This process is shaped by a variety of forces – some internal, local, and statistical, such as Hebbian learning, and some keyed to external reinforcers, such as pleasure, pain, and the example or instruction of conspecifics. In either case, such learning consists in the gradual reconfiguration of one's synaptic matrices, that is, of the weights of the trillions of synaptic connections that allow one vector-implementing population of neurons to stimulate the next such population in the brain's hierarchy. Ultimately, it is the acquired character of these synaptic matrices that dictates the acquired dynamical structure of the activation space of the receiving population of neurons. In other words, these matrices dictate the acquired hierarchy of categories into

---

[8] McCauley, *The Churchlands*, chaps. 10 and 11.

which that space has been partitioned. The acquisition of concepts, on this view, is thus something that requires the intricate and simultaneous tuning of trillions of synaptic connections – the individual "coefficients" of the relevant matrix.

Conceivably, these synapses might be fixed genetically, as concept nativism will be bound to claim against concept empiricism. But it is difficult to see how we could have very much in the way of innate concepts, for that would require the genetic specification of the individual weight-values – each one different – of many trillions of distinct synapses. The difficulty here is starkly arithmetic: there are roughly 100 trillion or 10^{14} synaptic connections in a normal human brain, but there are only about 10 billion or 10^9 functional base-pairs or "letters" in the human genome. Genetic information is wonderfully compressed, of course, but sheer noise *cannot* be compressed, and, as F&L themselves have correctly urged on me, every human brain is utterly unique in its connectivity at the synaptic level. We have no analog here for the "ten fingers, two eyes, one spinal column" sorts of features that every normal human *shares* and that thus admit of a genetic compression that can reliably be read out the *same way* in every normal fetus. On the contrary, synaptogenesis, at the level of the placements and the weights of individual synapses, is a profoundly idiosyncratic process. Like the correlation between the precise positions of the respective leaves within the outer canopies of two roughly spherical oak trees, the correlation between the precise placement and values of your synapses and the placement and values of mine is little different from zero. Each is a scatter. At that level of analysis, there is no correlation. From the point of view of the genome – which *is* almost identical across individuals – those details are the sheerest noise.

Accordingly, and with the possibility of information *compression* for synaptic weights generally thus put aside, even if the entire genome were somehow devoted to the specification of synaptic weights, and at the improbably generous rate of only one base-pair per synapse, the genome must still fall short of the information-storage capacity required by at least five orders of magnitude. In fact, the situation is even worse than this, because the functional unit here is not the single base-pair of nucleotides. It is the *sequence* of base-pairs adequate to construct a specific protein, and such sequences are typically 10^3 base-pairs or longer. Barring information compression once more (random structures, recall, are incompressible), the genome must therefore fall short of the capacity required by at least *eight* orders of magnitude. Concepts, it seems, *must* be learned from the environment, at least if they are globally embodied in the well-tuned microconfigurations of our 10^{14} synapses.[9]

Equally important, to their being learned, is *what* is learned, and here I put aside the issue of nativism versus empiricism to address the issue of atomism versus holism. If a typical three-layer network is successfully trained for some discrimination task, such as recognizing the gender, family, or individual identities of various faces, then it learns far more than just a reliable set of F&L-style causal responses to the real-world features at issue. On the contrary, the acquired structure of the network's activation space – that is, the metrically related family of prototype points – contains systematic and detailed *information*. Specifically, it contains information about such things as the structure of individual faces; about the general contrasts between the structures of male and female faces; about the various differences and similarities that divide and unite the face images into genetic families; and even about some set-inclusion and set-exclusion relations, such as the fact that Mary, Janet, and Liz are all females (all three individual prototype points lie within the female subspace), and the fact that no females are males (the male and female activation subspaces are nonoverlapping).

A clear example of the global informational richness typically acquired is Cottrell's original face-recognition network,[10] which revealed its acquired wisdom when fed, at the input layer, *incomplete* versions of the various faces it had previously been trained to discriminate. Figure 8.7a is an input photo of Mary, but with twenty percent of her portrait occluded by a gray "blindfold." The trained network identified her correctly even so.

Subsequent experiments revealed that the network's vectorial response to that attenuated input, at the all-important middle population of neuronlike units, is a response that correctly represents the *entire*

---

[9] This is the most fundamental objection against concept nativism of which I am aware. But it is just one member of a large family of negative considerations. For a recent and most welcome evaluation, see J. Elman et al., *Rethinking Innateness: A Connectionist Perspective on Development* (Cambridge, MA: MIT Press, 1996). For some recent modeling of the dynamics of axo-dendritic arborization, see C. Cherniak, M. Changizi, and D. Kang, "Large-scale Optimization of Neuron Arbors," *Physical Review E* 59, no. 5 (1999): 6001–9. On their account of synaptogenesis, "DNA-based mechanisms do not seem to be required" (p. 6008).

[10] G. Cottrell, "Extracting Features from Faces Using Compression Networks: Face, Identity, Emotions, and Gender Recognition Using Holons," in D. Touretzky et al., eds., *Connectionist Models: Proceedings of the 1990 Summer School* (San Mateo, CA: Morgan Kaufmann, 1991): 328–37.

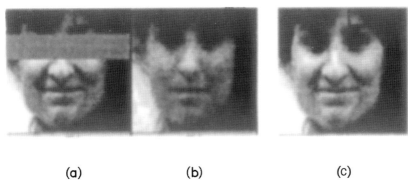

(a) **(b)** **(c)**

FIGURE 8.7. (*a*) An input photo of Mary with twenty percent of her face occluded. (*b*) The middle-layer representation of that occluded input, after decompression. (*c*) For comparison, an original training photo of Mary, unoccluded.

face of Mary. Figure 8.7*b* displays that middle-layer portrayal, after decompression. (For comparison, an original training photo of Mary is provided in Figure 8.7*c*.) The occluded portions of the degraded input are, at the middle layer, accurately filled in by the mature network. The filling is not perfect, but it is appropriate. Clearly, the network has managed, during training, to absorb and retain the information that faces in general have eyes, and, moreover, that Mary's face in particular has *these* sorts of eyes. And what it does for Mary, it does for all the other faces in its original training set. A partial input, within limits, will evoke at the middle layer a representation of the entire face, whichever face it might be.

Moreover, the narrow region of the abstract activation space that contains all of the Mary-vectors is *close to* the regions that contain the vectors of faces variously *similar* to Mary's. Similar faces are represented by proximate vectors, and dissimilar faces by comparatively distant vectors. Thus, all the female-face vectors are grouped together in one partition of the overall space, and all of the male-face vectors are grouped in a distinct partition. Moreover, mannish female faces and effeminate male faces will both be coded by vectors quite close to the hyperplane that divides those two partitions. The trained network, evidently, contains a good deal of information about both the concrete structures and the abstract relations within the domain of its sensory experience. That information is embodied in the high-dimensional and hierarchical set of similarity and dissimilarity relations that now configure the activation space of its middle layer. It is embodied, to use our earlier idiom, in a metrically determinate family of high-dimensional prototype points.

Nor is that acquired information an accidental accretion or an inessential luxury. The trained network is able to make the relevant set of perceptual discriminations because, and only because, it has acquired that systematic information. Fodor has spoken at length[11] on the lawlike connections that are supposed to bind our concepts to determinate features of the world, and bind them *independently* of any general beliefs or information that we might have about those features. But the idea, thus qualified, is implausible on its face. Human children learn readily to recognize faces, and dolls, and cookies, and socks. But there are no *laws of nature* that comprehend these things qua faces, dolls, cookies, and socks. All of them, to be sure, have causal effects on the sensory apparatus of humans, but those effects are diffuse, context dependent, high-dimensional, and very hard to distinguish as a class from the class of effects that arise from many other things.

This is why a student network has to struggle so mightily to assemble even a roughly reliable profile of diagnostic dimensions, a profile that will allow it to distinguish *socks* from shoes, boots, slippers, and sandals, not to mention mittens, gloves, tea cozies, and cloth hand-puppets. Having a reliable discriminatory response to socks is utterly dependent on this acquired command of a broad range of individually *in*adequate but overlapping and collectively trustworthy diagnostic dimensions. Those assembled dimensions make up the network's activation space, and their acquired sensitivities are what dictate whatever global similarity metric the space displays. Without such a structured family of activation-space categories – that is, without at least some understanding of the internal *character* of the feature being discriminated, and its various *relations* to other features whose character is also understood – the network will never be able to discriminate that feature from others in its perceptual environment. There are no laws of nature *adequate* to make the desired concept/world connections directly. The only access a human child can hope to have to the features cited, and to almost every other feature of conceivable interest to it, lies through a high-dimensional informational filter of the highly instructed and well-informed sort described. But such a filter already constitutes what Fodor portrays as unnecessary – a nontrivial weave of acquired *knowledge* about the various features in question. We are back, once again, to meaning holism.

Before leaving this point, let me emphasize that this is not just another argument for meaning holism. The present argument is aimed squarely

---

[11] *A Theory of Content and Other Essays* (Cambridge, MA: MIT Press, 1990).

at F&L in particular, in that the very kinds of causal connections they deem essential to *meaning* are in general impossible, save as they are *made* possible by the grace of the accumulated knit of background knowledge deemed essential to meaning by the holist. That alone is what makes subtle, complex, and deeply context-dependent features discriminable by any cognitive system. Indeed, it is worth suggesting that the selection pressure to make these ever-more penetrating discriminative responses to the environment is precisely what drove the evolutionary development of higher cognitive processes in the first place. Without such well-informed discriminative processes, we would be stuck at the cognitive level of the mercury column in a thermometer and the needle position of a voltmeter.

Beyond that trivial level, therefore, we should adopt it as a principle that there is "No Representation Without at least Some Comprehension." And the reason is not that we have bought into some a priori analysis of the notion of "representation." The reason is that, in general, representations cannot do their cognitive *job* – namely, allow us to make relevant and reliable discriminative responses to the environment and, subsequently, to steer our way through that environment – in an informational vacuum. If you are to have any hope of recognizing your situation within a complex environment, then you had better know a good deal about that environment.

This point also undermines Tiffany's deliberate neutralism about the ways in which SSS-style vehicles might get their semantic content. To generalize the point just made against F&L, no cognitive system could ever *have* the intricate kinds of causal, informational, or teleological sensitivities variously claimed for them, save by virtue of its possessing a systematic *knowledge* of the world's physical and causal structure. The embedding network of information so central to holism is functionally essential to any cognitive system above the level of an earthworm.[12]

At issue here is whether connectionism has anything to do with concepts and meaning. Let us summarize the points made so far. A structured activation space, embodying a family of prototype points, is a molar-level entity whose structure can be shared across physically diverse cognitive agents. It constitutes a background reservoir of systematic information

---

[12] After writing this section, I learned that Rob Cummins has urged the same general point in "The Lot of the Causal Theory of Mental Content," *Journal of Philosophy* 94, no. 10 (1997): 535–42. As he puts it there "distal" properties are "nontransducible." I concur. What is novel here is a peculiarly connectionist account of what that claim means, and of how any cognitive creature manages to transcend that barrier.

about the environment on which it was trained, about the principal features that the environment displays, and about a great many of the relations that hold between them. That internal system is acquired slowly, in the course of many interactions with that environment, and its acquired profile is sensitive to both the statistics of those interactions and the peculiar behavioral demands placed on the network. Once that structured space of possible patterns is in place, specific interactions with the environment will produce specific activation-patterns within that space, and those patterns are instrumental in producing the specific output behavior of the network, such as reliable discriminative responses to features of the environment.

Moreover, as we have seen, a network that possesses such a structured space can make successful discriminations of its learned categories despite partial or degraded sensory inputs, and its cognitive responses to such attenuated inputs contain defeasible information that goes substantially beyond what is strictly present in its inputs. That is to say, mere possession of the background framework constitutes knowledge of the world's general features, and specific activations within that general framework constitute ampliative representations ("recognitions") of the world's specific features here and now. If you didn't know that we are here talking about structured activation spaces, you could certainly be forgiven for mistakenly thinking, just for a moment, that we had been talking about conceptual frameworks.

VI. Prototype-Centered Regions as Individual Concepts

Do the parallels run any deeper than this? Yes they do. Using the vocabulary of the neural-network approach, we can also tell an illuminating story about individual concepts within the enveloping framework. To begin, a given concept encompasses a substantial *range* of distinct but closely related cases, in that the mature network will have generated, in the course of its learning the concept, a proprietary *volume* within its activation space, a volume that confines all of the possible points (i.e., neuronal activation vectors) that count as determinate cases of the abstract determinable that is the concept proper. The nontypical or marginal cases of the concept reside toward the periphery of that volume, and its center-of-gravity point represents a prototypical instance of the concept.

It is these prototype points that I have been leaning on, in the earlier sections of this paper, as the salient elements of any network's global conceptual framework. But it is really the comprehensive volumes, and the internal similarity metrics that structure them, that do the work. It is they that dictate the location and the character of the various prototype points, and it is they that capture most, if not all, of the instances the network actually encounters. A network may live out its entire life and never encounter, not even once, a perfectly prototypical instance of any of its categories. Accordingly, it may never produce an activation pattern at exactly the activation-space position of any of its internal prototype points. But it will be a conceptually competent network just the same.

Incidentally, this fact provides a plausible answer to Plato's classic question, "How can we have ideas of perfect or ideal *F*s when all we ever perceive are imperfect approximations to *F*s?" The answer is, "Because a broad sample of imperfect approximations is adequate to project a smooth metric that will capture all cases – the marginal, the prototypical, and even the hyperbolic cases."

This capacity to represent a *variety* of distinct perceptual cases as falling into the *same* encompassing conceptual volume also provides a vindication of Locke's notion of "abstract general ideas" over Berkeley's impatient objection that any "image" in the mind has to be entirely specific and particular in its properties. Perhaps Berkeley was right about images, but a concept is not an image. Rather, having-a-concept is having-the-*capacity* to represent each of a variety of relevantly related particular cases as lying within the same narrowly confined subvolume of activation space, a subvolume that bears a relatively fixed set of distance relations to a great many other such preferred subvolumes. Crudely speaking, a concept is not an image, but an isolated and graded range of *possible* images. And to *have* a concept is to *command* that well-informed range of possible representations.

The nonlinear metrics that typically characterize any concept's peculiar subvolume within activation-space also provide an explanation of so-called category effects.[13] This is the tendency of normal humans to make similarity judgments that group any two within-category items as being much closer together than any other two items, one of which is inside and one *outside* the familiar category. Humans do this even when, by any "objective measure" of sensory inputs, the similarity distances are the same.

---

[13] S. Rosen, and P. Howell, "Auditory, Articulatory, and Learning Explanations of Categorical Perception in Speech," in S. Harnad, ed., *Categorical Perception: The Groundwork of Cognition* (Cambridge: Cambridge University Press, 1987), 113–60. See also Harnad's useful introduction, 1–4.

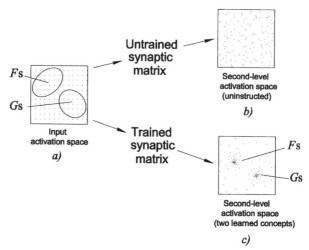

FIGURE 8.8. (*a*) The space of possible input vectors for a simple network. (*b*) The corresponding activation vectors produced at the network's middle layer, after processing by a matrix with random coefficients. (*c*) The corresponding activation vectors produced at the network's middle layer, after processing by a matrix with coefficients trained to discriminate *F*s from *G*s.

Typical feedforward networks (which deploy nonlinear squashing functions to mimic the response behavior of biological neurons) show the same sorts of nonuniform groupings when trained to discriminate a family of mutually exclusive categories. We can illustrate the general tendency with the cartoon example of Figure 8.8. The range of possible sensory patterns at the input layer of neurons is represented by the array of points within the two-dimensional activation space of Figure 8.8*a*. Each one of those many input patterns is projected, in sequence, through a matrix of *randomly* set synaptic connections, so as to produce a corresponding point in the activation space of the next layer of neurons, as shown in Figure 8.8*b*. The result is a noisy but more or less uniform distribution of points within that receiving space.

Contrast this case with the case of a network that has been successfully trained to discriminate between two perceptual categories, *F* and *G*. Here the matrix of synaptic connections has been carefully tuned so as to yield a very different distribution of reactive patterns at the second layer. As shown in Figure 8.8*c*, the resulting points now have a decidedly nonuniform distribution: the original sensory-layer points that fell within the class *F* have produced a family of second-layer activation-space points that are clustered very close together. The same is true for the input and the second-layer points associated with the class *G*. And between those two clusters lies a comparative no-man's-land. What were uniform similarity distances within the input stimulus space have become variously nonuniform distances within the "conceptual" space of the second-layer neurons. The boundaries of the classes *F* and *G* are no longer arbitrary lines dividing a uniform space, as in Figure 8.8*a*; they are now dynamically marked by relative deformations in the similarity metric, as in Figure 8.8*c*. Hence the empirical profile of human similarity judgments that prompted this discussion.

We can also see, once more, the significance of the expression, "prototype points" – they are the focal points of the sorts of activation-point clusterings visible in Figure 8.8*c*. Since they are the focal points of the learned distortions in the second layer's similarity space, their locations can be identified independently of determining what, if anything, they may represent. The nonuniform metric of that space can be calculated from the configuration of the assembled synaptic weights that project to it (recall, it is the latter that determine the former). Alternatively, one can determine the deformed metric experimentally, by entering a large but random series of input vectors at the sensory layer (a set that fairly samples the space of *possible* input vectors) and then observing what clusterings emerge at the second layer. This is what is portrayed in the shift from Figure 8.8*a* to Figure 8.8*b*, and from Figure 8.8*a* to Figure 8.8*c*, respectively. But the space of Figure 8.8*b* is uninstructed, whereas the space of Figure 8.8*c*, thanks to the matrix of synapses that shapes it, has plainly learned two distinct concepts.

One thing that concepts do *not* do, on our view, and in contrast to historical accounts such as Locke's and Hume's, is form compositional hierarchies such that all complex concepts are literally constituted, by suitable concatenation and recursion, from a finite lexicon of simple concepts. This historical view implies a unique *de*composition of any complex concept into a determinate set of simple concepts, so that anyone who possesses the relevant complex concept must ipso facto possess as well the several simples from which it is made.

The architectural appeal of such a "molecular" view is obvious, and a mature conceptual framework does indeed display a hierarchical organization of some sort. But developmental facts indicate that the classical view at issue cannot be *quite* right. Children learn to discriminate faces, from other things and from each other, substantially before they can do the same for eyes, noses, mouths, or ears. And they subsequently learn, in turn, to discriminate eyes, from other things and from each other, substantially before they can do the same for pupils, eyelashes, irises, or lenses. "Entry-level" or "basic-level" concepts – those that children learn first – are seldom if ever the so-called simple concepts favored by Locke and Hume.[14] In general, the first-learned concepts are what those historical authors would have called highly "complex" ideas, such as *cookie, dog, face, bird,* and *shoe.* Only later do children begin to acquire a family of *sub*ordinate concepts, such as *robin, sparrow,* and *crow* to fine-tune their antecedent concept *bird*; or *spaniel, lab,* and *husky* to fine-tune their antecedent concept *dog.* And even more slowly do they develop a framework of *super*ordinate concepts, such as *animal* to unite birds, dogs, and horses, or *clothing* to unite shoes, socks, and pants.

Evidently, the orderly assembly of primary semantic atoms into secondary semantic molecules is not the developmental pattern displayed here. On the contrary, the hierarchical structure that does emerge reflects the child's gradual learning of *the world's objective structure* more than it reflects the gradual compositional articulation of some innate lexicon of conceptual simples. And that is as it should be. Since there is no analytic/synthetic distinction,[15] acquiring a system of meanings can be nothing less than acquiring a body of presumptive knowledge about the world.

To be sure, the classical view can afford to concede that "simple" concepts are often "first activated" in compositional concert, so as to respond appropriately to commonly perceived complexes in the world. In this way, perhaps, the developmental facts about children might be rendered less awkward. But it is precisely that compositional assumption that is here being questioned. The recent lesson from the external behavior and the internal organization of artificial neural networks is that the expected forms of compositional structure are quite absent, and quite unnecessary. A network, such as Cottrell's network (Figure 8.9*a*), that is entirely competent at discriminating human faces from other things, need have no capacity at all to discriminate such "simpler" things as isolated noses, eyes, mouths, or ears, and no identifiable neuronal axes or internal subvolumes of activation-space that have anything to do specifically with noses, eyes, mouths, or ears. Bluntly, it has the "complex" concept *face*, but it does not yet have any of the "simple" concepts that common sense and the Hume/Locke tradition might presume to constitute it.

---

[14] J. M. Anglin, *Word, Object, and Conceptual Development* (New York: Norton, 1977), 220–8, 252–63.

[15] W. V. Quine, "Two Dogmas of Empiricism," in *From a Logical Point of View* (New York: Harper and Row, 1963), 20–46.

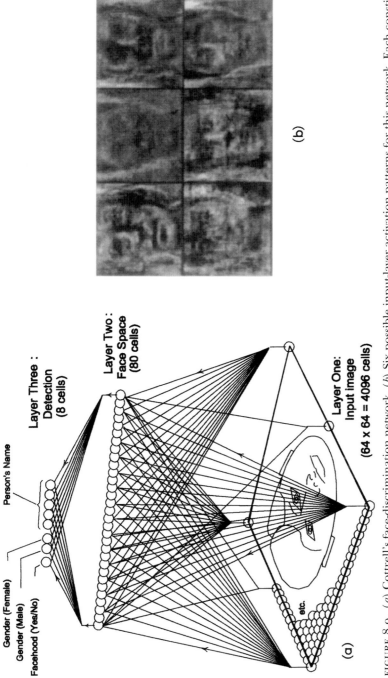

(b)

FIGURE 8.9. (a) Cottrell's face-discrimination network. (b) Six possible input-layer activation patterns for this network. Each constitutes the "preferred stimulus" of exactly one of the eighty middle-layer neurons. Each serves as one of eighty "templates" to which any input image is "compared," so that each input image receives a highly individual, eighty-dimensional "profile" of middle-layer activations.

What the network does have, instead, is eighty neurons at its second layer, each of which has a proprietary but vague, noisy, and retina-encompassing input pattern that is its "preferred stimulus." That proprietary pattern, when delivered as input at the retinal layer, produces the maximal level of excitation or activation in that second-layer neuron. The distinct preferred-input pattern for each of those second-layer neurons need not be, and almost certainly will not be, a pattern that the network as a whole has ever encountered. But the real input patterns that each neuron actually does encounter will fall variously close to or far away from approximating its preferred pattern, and thus its individual activation level will be a measure of how closely any given input image overlaps or resembles that preferred pattern.

With eighty neurons, each of which performs its own discrete variation on this similarity-measuring theme, the network's second layer yields a unique eighty-dimensional assessment of any image presented to the input layer. What is welcome, in the successfully trained network, is that *all and only faces* will pass this diffuse and high-dimensional assessment. And it is these hard-won global assessments, embodied at the second layer, that the network's final or detection layer is tuned to discriminate in turn.

Here it must be emphasized that the various "preferred stimulus" patterns peculiar to each second-layer neuron typically do *not* correspond to anything that common sense would regard as a conceptual "simple." (See Figure 8.9*b* for the actual preferred stimulus empirically determined for each of six neurons in Cottrell's face network.) For one thing, those patterns are highly complex and span the entire retinal surface. Hence we have the term "holons," coined precisely to capture this feature.

Additionally, the preferred stimulus for any given neuron at the second layer of Cottrell's network is an intricately ordered n-tuple with $n =$ fully 4,096 precisely placed constituting elements. (The network's input layer or "retina," recall, has $64 \times 64 = 4,096$ pixels.) These patterns are not "simples" at all, especially because most of them are *already* elusively facelike in their global organization.

Finally, there are far *too many* such patterns – one for each and every neuron at the second layer – plausibly to play the role of innate simples. That would require fully eighty simples even for Cottrell's comparatively tiny artificial face-recognition network, and it would require something like 10^8 or 10^9 "simples" for the human face-recognition network alone, whose relevant neuronal population is correspondingly larger. This signals a *reductio* of the idea that the neuronal axes of a neural network might correspond to the "simple" ideas of the traditional compositional story. Hume would have expected the concept of *face*, for example, to decompose into something like ten simpler concepts, perhaps a hundred, conceivably a thousand. But a hundred million, or a billion, component concepts? This is no longer what Hume, or F&L for that matter, had in mind.

The preferred stimuli of the second-layer neurons of any network are sometimes referred to as the "microfeatures" to which those neurons have become sensitive. And this may tempt us to assimilate them to the hoped-for classical simples. But it won't wash. First, in real brains, these preferred stimuli are more accurately characterized as *nano*-features than as deci-, milli-, or micro-features. Second, each one is rich with internal structure anyway. And finally, they do not concatenate in anything like the Boolean fashion that the classical story requires. Instead, it is a mere statistical preponderance of those nanofeatures, suitably distributed across the coding population, that drives the network's various output responses.

(The misconceived assimilation here abjured is briefly ascribed to me by F&L,[16] but that is a mistake as well. The three-dimensional facial state-space whose axes are labeled "nose width," "eye separation," and "mouth fullness" – F&L's presumptive "smoking-gun" Figure 1 – was deployed by me in my 1995 book[17] for two reasons: First, to introduce the general notion of state-space representations to a naïve audience; and second, subsequently to *contrast* that classically labeled example with the story of what happens in real networks such as Cottrell's, where the preferred stimulus for each neuronal axis at the second layer turns out, empirically, to have a character utterly unlike the classical features of the introductory example. The discussion of this vital contrast takes up pages 46–9 of that book. F&L's slip here – however natural – reflects a persistent and widespread misconception of how a trained neural network represents the world.)

In light of all this, it should come as no surprise that networks can and regularly do have "complex" concepts *without* also having any classical simples as their functional constituents. For on our view, *all* concepts are complex. All of them have an intricate internal structure, a structure with a dimensionality equal to the number of neurons in the space that embeds them. But little or none of that structure is usefully captured in

---

[16] Fodor and Lepore, 391, 399.

[17] P. M. Churchland, *The Engine of Reason, Seat of the Soul: A Philosphical Journey into the Brain* (Cambridge, MA: MIT Press, 1995), 28.

the classical compositional story. And the various "preferred stimuli" for the neurons of any human are almost certainly learned and idiosyncratic to each individual, rather than innate and standard across every member of the species.

This means that the internal structure of any given concept is likely to be highly variable across distinct individuals. What is highly *similar* across individuals is the framework of activation-space *distance relations* that structure each individual's conceptual framework as a whole. This can indeed yield a substantial number of classical-style inclusion relations (for example, in Cottrell's face-recognition network, all female faces are faces), but we need no longer be bound by the picture of "complex" ideas being literally and invariably compounded from preexistent "simple" ideas.

In an earlier paper,[18] Fodor et al. wisely reach a similar conclusion against the classical idea of a compositional hierarchy of concepts, but the lesson drawn therefrom, by Fodor, is that essentially *all* concepts are thus without internal structure and must therefore be innate. The starting place is good, but both inferences are extravagant. As we have seen, there are more ways of having internal structure than are recognized by the classical story. And as we have calculated, the chances of that structure's being innate – that is, the chances of its being coded in the human genome – are indistinguishable from zero. Fodor's insight concerning the absence of classical definitional structures is to be applauded. But we can here see our way clear to a quite different explanation of why they are absent. The SSS account, of what concepts are, successfully reconstructs the actual hierarchies that conceptual frameworks do display, but it positively *refuses* reconstruction of the classical Locke/Hume picture.

In sum, activation-space regions with suitably tuned similarity metrics are prime candidates to explicate the notion of a conceptual framework, particularly if one is a semantic holist. And, as we have seen, there are compelling grounds for being a semantic holist.

I close this section by remarking that the theoretical position here articulated has a history of some length and an experimental tradition of some vitality.[19] Shepard (1968) suggests explicitly that what matters

---

[18] J. A. Fodor et al., "Against Definitions," *Cognition* 8 (Amsterdam: Elsevier Science, 1985). Reprinted in E. Margolis and S. Laurence, *Concepts: Core Readings* (Cambridge, MA: MIT Press, 1999), 491–512.

[19] R. N. Shepard, "Cognitive Psychology: A Review of the Book by Ulrich Neisser," *American Journal of Psychology* 81 (1968): 285–9; also, "Multidimensional Scaling, Tree-Fitting, and Clustering," *Science* 210 (1980): 390–7; W. V. Quine, "Natural Kinds," *Ontological*

for representation is a global or 'second-order' isomorphism between an entire *family* of concepts and the entire *range* of objects or features that they represent, as opposed to any 'first-order' isomorphisms between concepts and objects taken singly. Quine (1969) and Goodman (1972) have both explored "similarity spaces" in an attempt to understand the nature of our perceptual categories. Kuhn (1974) deploys the notion in an early attempt to explicate his idea of a conceptual paradigm. Churchland (1989) deploys the same notion to explicate the nature of theories, explanation, and two quite different kinds of learning. More recently, Shimon Edelman's (1998) paper reports on the cognitive behavior of an artificial network, *Chorus*, that deploys a smallish population of distinct "feature detectors" to create a collective activation space that also embodies a similarity metric and a population of prototype regions, as in the several networks discussed earlier. I commend all of these papers to the reader's attention, especially the one by Edelman, who provides an insightful philosophical commentary (occasionally opposed to my own) to accompany the experimental and theoretical psychology.

VII. The Portrayal of Worlds

But what about *intentionality*? How can a purely internalist account of sameness-of-meaning hope to account for the "pointing beyond itself" that is traditionally seen as the hallmark of concepts? Fair questions. To which we can propose some contentious but not entirely unfamiliar answers. To begin, there is no problem in principle here for our internalism, as a cartoon analogy will illustrate. If you create on a computer, using a suitable drafting program, an architectural drawing of a proposed house, and then print out two copies of the relevant drawing, it will be no surprise that a purely internalist criterion will allow you to say that each printout is an instance of the same picture, that each presents *the same portrayal* of the proposed house (or the actual house, should it happen to be built). For each piece of paper contains a system of points and lines whose respective internal positions and lengths are identical.

Relativity and Other Essays (New York: Columbia University Press, 1969), 69–90; N. Goodman, *Problems and Projects: Seven Strictures on Similarity* (Indianapolis, IN: Bobbs-Merrill, 1972); T. S. Kuhn, "Second Thoughts on Paradigms," in F. Suppe, ed., *The Structure of Scientific Theories* (Urbana: University of Illinois Press, 1974); P. M. Churchland, *A Neurocomputational Perspective* (Cambridge, MA: MIT Press, 1989), chaps. 9–11; S. Edelman, "Representation Is Representation of Similarities," *Behavioral and Brain Sciences* 21 (1998): 449–98.

(Note also, recalling two issues from Section II, that global rotations on the page do not matter here either, and neither would any rotational self-symmetries of the house portrayed.) Indeed, if it is *the character of the portrayals* involved that concerns us, then internalist criteria for sameness and differences are clearly the appropriate ones. After all, the house has not been built yet, and may never be. The internalist character of our *identity* criterion is therefore not a problem for me. My problem lies elsewhere. It lies in telling a nonvacuous story about how concepts *portray* the world, and about how such portrayals can subserve the many practical skills we display.

On the view of concepts defended in the preceding sections, an individual's background conceptual framework already constitutes a portrayal of the world's general features – roughly, those features stable over time – while specific activations of that background machinery typically constitute specific portrayals of the world's local character here and now. But what sense(s) of "portrayal" are we confronting here? If, in a Quinean or Davidsonian spirit, we were to construe a conceptual framework as a network of general sentences accepted by its possessor, then we could appeal to the familiar notions of reference, extension, set inclusion, logical structure, and recursively reckoned truth, all in hopes of explaining how it is that such a framework can portray a world, either accurately or inaccurately. But that is not the construal of a conceptual framework that we have been exploring in this essay. Our construal addresses conceptual frameworks as decidedly *sub*linguistic entities. Accordingly, *all* of that beloved logical machinery is here denied us, as a potential *explanans*. The way in which overtly linguistic structures represent the world (if they do) is something that itself stands in need of explanation, one properly grounded in *sub*linguistic terms.

Where, then, should one start? Why not with the brain? The brain has a great variety of representational systems – presumably, one for each anatomically distinct neuronal activation-space. The representations that concern us here are the enduring or abeyant ones, as opposed to the fleeting activation-patterns – occasioned by sensory activity, for example – that take place within the comparatively stable background representational framework. We are here concerned, that is, with the lasting system of prototype points, and with the similarity and difference relations that structure the activation-space that embeds them (see again Figure 8.2), as opposed to any momentary activation patterns within that space. What we want to know is how, or in what sense, do such structured activation-spaces *portray* the world?

Let me approach a general answer to this question by way of a nonbrain analogy. Many of us have by now encountered, if only in an upscale rental car, a GPS (Global Positioning System) display. The rental car contains a dashboard-screen display of a small portion – perhaps six or seven square blocks – of a vast, computer-stored street map of the surrounding urban environment. A stored street map is of course a useful resource in its own right, but the onboard guidance system at issue offers a new twist. It is electronically linked to a set of geostationary GPS satellites that tell the onboard system exactly *where* on Earth's surface the car happens to be at any moment. The onboard computer then displays, to the driver, the tiny portion of the larger urban map that currently contains the car. That dashboard display has a small, upward-pointing icon of the car at its focal crosshairs, and the display is appropriately oriented, relative to that icon, to represent the current direction of travel. As one drives around town, one can thus observe a detailed street map flowing by the displayed car icon, as if one were watching one's car from an overhead helicopter in close pursuit.

This opening analogy is deployed for several reasons. First, it highlights the contrast between the stored urban map as a whole, and the momentary, punctate, mutually exclusive, and constantly changing *locations* on the map that get sequentially displayed on the car's dashboard (these are the "contents of sensory experience," as it were). Second, it highlights the fact that what makes that stored map a *portrayal* or *representation* of the entire urban area is the usual *relation-preserving, abstract, projective mapping* that makes any map a map. In particular, that stored map is not a successful and potentially useful map because of any *causal* relations that it happens to enjoy to the external world. The GPS link, for example, is utterly inessential to its maphood. The system would work just as well if it were keyed instead to bar-coded magnetic beacons embedded in the roads every fifty feet; or to a video system, on the hood, trained to read street signs; or to an inertial guidance system, indexed once at the factory. Indeed, the stored map would remain a perfectly legitimate map even if the car were to lose any and *all* of its "sensory" access to its environment. It is a map because of its own internal structure, and because there is an abstract, relation-preserving mapping from the global street system surround to that stored internal structure. And it is precisely because of its independent status in that regard that the map can, when given an appropriate but essentially arbitrary "sensory system" like the GPS link, be *used* in the fashion described – as a means of displaying one's current position in the space of "locational possibilities," as a means of anticipating its local features (a park to the right, an ocean view to the left), and as a navigational guide toward other places not yet occupied.

Third, the analogy highlights the possibility of both fleeting and lasting *errors* of representation. The map itself can be a good or a bad map in many dimensions. And the sensory mechanisms for activating local bits of the global map may also display occasional glitches, as when the dashboard display wrongly portrays the car as being halfway across San Diego's Coronado Bridge when the car is actually in downtown La Jolla. These latter failures, note, need imply no fault in the background map itself, only in its local application at a specific time.

Fourth, the analogy suggests, for different vehicles with different purposes, a great *variety* of possible stored maps, each with its own representational virtues and inadequacies. An internal map might concern itself with street paths, as in the car example at issue. But if the map is stored inside a TV-news helicopter's computer, to guide the aircraft's navigation, then the stored map might focus instead on the landscape's topographical features; on the location and altitude of obvious mountain peaks or other navigational hazards, such as local broadcast towers and skyscrapers; and on the location of local airstrips and their always-busy takeoff and landing paths. Alternatively, if the helicopter belongs to a city's maintenance department, its stored map might portray the complete grid of the city's underground water-main, storm-drain, and sewer systems. Or, if the vehicle is a hostile alien fighter-bomber aircraft, its stored map (still GPS-activated) might represent San Diego's system of acquisition radars, the locations of the city's naval and marine antiaircraft missile batteries, and the locations of this week's targets. And so on.

What we are seeing here is the entirely real possibility that the very *same* "sensory system" or "causal link" to the external world (a GPS link, for example) can opportunistically drive, activate, or serve the deployment of any one of a great many quite *different* sorts of stored internal maps. And that is because the maps can be identified as this, that, or the other sort of map *independently* of whatever causal connections they may (or may not) have to the external world. It is the map's internal structure that makes it the specific portrayal that it is. And it is the existence of an abstract, relation-preserving, projective mapping, from some external domain to that map, that makes it a good or an accurate portrayal of that external domain. Causal connections enter the picture only if, and only when, the map is finally put to some use or other.

This is ultimately why causal accounts of semantic identity and semantic similarity, such as F&L's, are doomed to failure. They cannot tolerate, acknowledge, or hope to explain the possibility that very *different* "systems of meaning" (i.e., different conceptual frameworks) can be causally activated by the very *same* causal factors in the external world. For on all such accounts, it is precisely those external causal factors that fix or determine the semantic identity of the internal concepts that they activate. "Same external causes, same meaning." On a causal account, therefore, the *meaning* of the dashboard display in any of the four vehicles described must always be nothing other than, "You are at such-and-such a position relative to the three triangulating GPS satellites." But that just isn't so. Position relative to the satellites is part of the causal story of how the four different maps get usefully deployed in real time, but it is not what each, or any, of those maps means. The four maps mean four different things, each tailored to the peculiar concerns of its user.

We have known for some time that meanings transcend both the sensory systems and the external causes that may occasionally activate them. Churchland 1979,[20] for example, explores at some length the possibility that our native and unmodified sensory systems can serve to drive (i.e., make systematic activations within) a variety of very different conceptual frameworks, depending on the details of one's training and education. What we have before us, in the present paper, is an updated, neurocomputational account of what concepts are and of how their semantic identities are specified, an account that allows for this "thousand-flowers" possibility. Any account that precludes that possibility – F&L's, for example – is a nonstarter.

Having leaned so hard on this extended analogy with maps, I am obliged to address the issue of how apt this analogy may be. Is a conceptual framework, even on the neural accounting outlined earlier, really like a map? Certainly not, if we conceive of a map simply, as a two-dimensional representation of a two-dimensional family of objective spatial points and distances.

But suppose we conceive of a map more broadly, as an n-dimensional structure, of arbitrary physical makeup, whose internal elements and structural relations mirror the elements and relations within some n-dimensional objective domain, where $n >> 2$? If we adopt this more inclusive conception, then the eighty-dimensional middle-layer activation-space of Cottrell's face-recognition network (see again Figure 8.9a)

---

[20] P. M. Churchland, *Scientific Realism and the Plasticity of Mind* (Cambridge: Cambridge University Press, 1979), chap. 2. The conceptual alternatives explored there are drawn from contemporary physical theory.

emerges as one instance of a map. It is not a map, as in the rental-car case, of possible *physical positions in geographical space*; it is a highly structured map, instead, of the range of possible *human faces*.

Equally, the activation space of our assembled motor neurons is presumably a map of the many limb positions possible for the human body – or, more likely, of the set of possible *motor sequences* possible for our muscular and skeletal systems. Similarly, and more simply, the structured activation space of our own color-opponent neurons in the LGN or V4 is a (somewhat problematic) map of the range of possible *objective electromagnetic reflectance profiles*. And so, I suggest, is *every* structured neuronal activation- space within the brain an abstract map of *some* objective domain of features, structures, or processes. All of those maps serve, as does the map in the GPS-equipped rental car, to inform our grasp and guide our navigation of the world at large. But they allow us to navigate far more than mere paths on Earth's surface. Collectively, they allow us to navigate abstract "paths" through social space, color space, thermal space, financial space, auditory space, limb-configuration space, moving-body space, and the high-dimensional space of causal processes generally. The mapping relations involved will often be complex, and sometimes *very* complex. But there need be nothing mysterious about them. They are all relation-preserving mappings, of some sort or other, from aspects of the world to learned structures in high-dimensional neuronal-activation spaces.

No doubt the relation-preserving mapping will be different from space to space. That is to say, no *single* projection relation emerges as *the* "secret of intentionality." For why ever should the brain confine itself to a single projection relation? Why should it not be as opportunistic as can be, and embrace any coding strategy, any projection relation, any mapping function, that will allow it to discriminate and anticipate important aspects of the world, and thus to navigate it in a more informed and successful fashion? It would seem only reasonable. And on this assumption, we would expect evolutionary time, and the enormous diversity of features in need of representation, to produce brains that display an intricate knit of diverse coding strategies, exploiting whatever projection relations happen to be both available and appropriate for the spatial, thermal, chemical, electromagnetic, mechanical, biological, and social realities that they must confront. Research on the nature of "intentionality," therefore, must be a piecemeal and empirical affair, undertaken anew for each matrix of synapses and each neuronal population in the brain. It must be an affair that expects to find a *proprietary* projection relation for each cognitive subsystem that is addressed. In this way, the brain's relation(s) to the world need not remain a mystery, and certainly not a *single* mystery.

In sum, the SSS account of what concepts are entails that a framework of concepts constitutes an abstract *picture* of some part, slice, or aspect of the objective world. It is not a "logical" picture, as the early Wittgenstein would have it – that is much too narrow. Nor is it a literal picture, as in a two-dimensional colored surface – that also is much too narrow. Rather, it is a picture in the more abstract sense that it is a complex physical structure whose functionally salient internal relations mirror the family of relations that make up the external domain portrayed. More strictly, there exists a relation-preserving mapping from the external domain to the acquired structure of the relevant neuronal-activation space.

In contrast to F&L's account, this account has the further advantage that it allows us to address the questions of how *accurate* or *inaccurate* a given conceptual framework might be, and of how *superficial* or how *penetrating* it is. This account also allows us to explain both the behavioral successes and the behavioral failures of the creature using that framework, because that internal world-portrayal is the creature's principal guide to the production of complex behavior, that is, to the creature's ongoing interaction with the world portrayed. Its chronic representational shortcomings and/or its occasional misapplications will serve to explain the behavioral missteps of its user. And all of this semantic and normative richness is in place before the phenomenon of language ever enters the stage.

Finally, what *about* language? It is a marvelous achievement, and it has launched us on a path so far denied to the other animals. How has it done this? And where is language going to fit into the larger cognitive picture sketched in this essay?

Perhaps as follows. Think of language not so much as a system for representing the world, but as an acquired *skill*, both a motor skill and a perceptual skill. But do not think of it as the skill of producing and recognizing strings of words. Think of it instead as the acquired skill of perceiving (opaquely, to be sure) and manipulating (again, opaquely) the brain activities of your conspecifics, and of being perceptually competent, in turn, to be the subject of reciprocal brain manipulation. We do not usually think of a dinner-table conversation in these terms, but evidently that is what is going on. I am both following and steering your own cognitive activities, as you are both following and steering mine.

Wittgenstein, perhaps, was halfway to this perspective when he insisted on the directive functions of language. Perhaps J. L. Austin was also, with his intricate taxonomy of performative utterances. I won't press the point, however. Certainly, neither of these thinkers had any ideas about the manipulation of brain activity in particular. Still, the widespread ability to monitor and manipulate the brain activities of one's conspecifics would evidently unite us, cognitively, as in no other species. Our cognition would thus occasionally become a *collective* activity, on a minute-by-minute and even a second-by-second basis.

Such an ability would provide a major advantage in that respect alone. And a secondary advantage would arise from the inevitable compounding of that initial investment. Over generations, the evolving *form* of that manipulational skill would itself come to embody useful general information, information transmittable from generation to generation as the skill itself gets passed down. Such a system would be wonderful. And it is. But it needn't constitute the basic machinery of cognition itself. *That* machinery is hundreds of millions of years old, and we share it with countless other species. The machinery of language, by contrast, is ours alone, and it is no older than we are. A theory of cognition must respect that fact.

9

Chimerical Colors

Some Phenomenological Predictions
from Cognitive Neuroscience

Abstract: The Hurvich-Jameson (H-J) opponent-process network offers a familiar account of the empirical structure of the phenomenological color space for
humans, an account with a number of predictive and explanatory virtues. Its
successes form the bulk of the existing reasons for suggesting a strict identity
between our various color sensations on the one hand, and our various coding
vectors across the color-opponent neurons in our primary visual pathways on
the other. But antireductionists standardly complain that the systematic parallels
discovered by the H-J network are just empirical correspondences, constructed
post facto, with no predictive or explanatory purchase on the intrinsic characters of qualia proper. The present paper disputes that complaint, by illustrating
that the H-J model yields some novel and unappreciated predictions, and some
novel and unappreciated explanations, concerning the qualitative characters of
a considerable variety of color sensations possible for human experience, color
sensations that normal people have almost certainly never had before, color sensations whose accurate descriptions in ordinary language appear semantically
ill-formed or even self-contradictory. Specifically, these 'impossible' color sensations are activation-vectors (across our opponent-process neurons) that lie inside
the space of neuronally possible activation vectors, but outside the central 'color
spindle' that confines the familiar range of sensations for possible objective colors. These extraspindle chimerical-color sensations correspond to no color that
you will ever see objectively displayed on a physical object. But the H-J model both
predicts their existence and explains their highly anomalous qualitative characters in some detail. It also suggests how to produce these rogue sensations by a
simple procedure made available in the latter half of this paper. The relevant
color plates will allow you to savor these sensations for yourself.

I. Introduction

The qualitative character of subjective experience is often claimed to
be beyond the predictive or explanatory powers of any physical theory

(Huxley 1866; Nagel 1974; Jackson 1982; Levine 1983; Chalmers 1996). Almost equally often, conclusions are then drawn concerning the physical irreducibility and the metaphysical distinctness of the subjective qualia at issue. Resistance to such dualist themes has typically focused on the dubious legitimacy of the inference just displayed (Churchland 1985, 1996c; Bickle 1998). The present essay, by contrast, focuses on the premise from which the inference is drawn. My burden here is to show that this premise is false.

I will illustrate its falsity by drawing a number of novel, counterintuitive, and, in some cases, patently paradoxical predictions concerning the qualitative character of certain highly unusual visual sensations, sensations produced under some highly unusual visual circumstances, sensations you have probably never had before. These predictions will be drawn, in a standard and unproblematic way, from the assumptions of what deserves to be called the Standard Model of how color is processed and represented within the human brain (Hurvich 1981; Hardin 1988; Clark 1993). I am thus posing only as a consumer of existing cognitive neuroscience, not as an advocate of new theory. But standard or not, this familiar 'color-opponency' theory of chromatic information processing has some unexpected and unappreciated consequences concerning the full range of neuronal activity possible, in an extreme, for the human visual system. From there, one needs only the tentative additional assumption of a systematic identity between *neuronal coding vectors* on the one hand, and *subjective color qualia* on the other – a highly specific material assumption in the spirit of the classical identity theory, and in the spirit of intertheoretic reductions generally – to formally derive the unexpected but qualia-specific predictions at issue.

Accordingly, these predictions provide no less than an empirical test of the identity theory itself, in one of its many possible (physically specific) guises. We may therefore approach with interest the question of whether the weird predictions promised earlier actually accord with the data of subjective experience, in addition to the question of whether and how those predictions arise in the first place. The several color plates provided with this essay, plus some experimental procedures to be described as we proceed, will allow you to test the relevant predictions for yourself. The aim is to produce in you color sensations that you have (almost certainly) never experienced before, sensations whose highly specific descriptions in commonsense terms are, by prior semantic lights, flatly self-contradictory. Nonetheless, those nonstandard sensations are real,

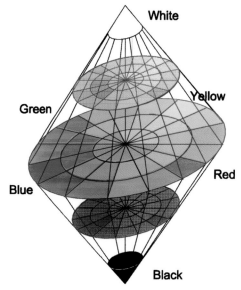

FIGURE 9.1. The classical color space.

their paradoxical descriptions are accurate, and the Standard Model predicts them all, right out of the box.

II. The Standard Model: The Color Spindle
and the Hurvich-Jameson Net

The many colors perceivable by humans bear a complex set of similarity
and dissimilarity relations that collectively position each color uniquely
within a continuous manifold. The global structure of that manifold has
been known since Munsell first pieced it together over a century ago. (See
Figure 9.1 for a slightly oversimplified rendition. A more accurate rendition would have both cones bulging outward somewhat.)[1] The agreeably

---

[1] Strictly speaking, Munsell intended his solid to represent the relations between the various
external or *objective* colors. But it serves equally well as a representation of the similarity-
and-difference relations between our internal color *sensations* as well. That is the use to
which it is here being put. That is, the spindle-shaped solid represents our *phenomenological
color space*. Be advised, however, that it provides only a first-order model. Its internal metric
is suspect, and we may well need a *four*-dimensional space to capture *all* aspects of human
color perception. But those complexities lie safely beyond the specific concerns of this
paper. For a broad summary, see R. Kuehni, *Color Space and Its Divisions: Color Order from
Antiquity to the Present* (New York: Wiley 2003).

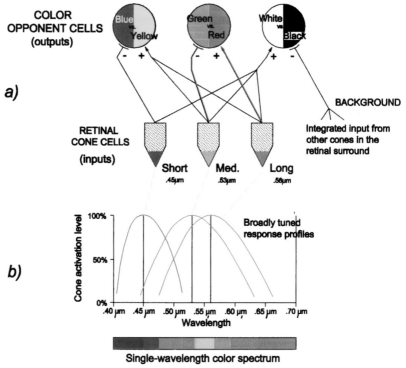

FIGURE 9.2. The Hurvich-Jameson network.

simple Hurvich-Jameson (H-J) net is a recent attempt to *explain* that global structure in terms of the known elements of the human visual system. It begins with the three types of cone-cells distributed across the retina, cells broadly tuned to three distinct regions of the visible spectrum, conventionally dubbed the short-, medium-, and long-wavelength (S, M, and L) cones, respectively. And it ends with three kinds of color-coding cells at its output layer, cells whose activity levels code for the simultaneous position of any visual stimulus along a blue-to-yellow axis, a green-to-red axis, and a white-to-black axis (see Figure 9.2).

On this model, the resting-levels of electrical activity in the three input cones are postulated to be zero, with a maximum level of 100%.[2] By

---

[2] As an aside, human cone cells respond to light with smoothly varying *graded* potentials (voltage coding), rather than with the varying frequencies of spiking activity (frequency coding) so common in the rest of the nervous system. This wrinkle is functionally irrelevant to the first-order model, which is why cellular activation levels are expressed *neutrally*, in what follows, as a simple percentage of maximum possible activation levels.

contrast, the default or resting-levels of electrical activity in the three second-rung output cells are postulated to be 50% of their maximum possible activation levels (the full range is zero spikes/sec to roughly 100 spikes/sec). Excursions above and below that midway default level are induced by whatever excitatory or inhibitory inputs happen to arrive from the various cone-cells below. Each output cell is thus the site of a tug-of-war between various input cones, or coalitions of input cones, working with and against each other to excite or to inhibit, above or below the spontaneous resting level of 50%, the particular output cell to which they severally project.

The so-called Green/Red cell is the simplest case, since its activation level registers the relative preponderance of the long-wavelength light over/under the medium-wavelength light arriving to the cones at the tiny area of the retina that contains them. A local preponderance of long over medium excites the L-cones more than the M-cones, which yields a net *stimulation* at the Green/Red cell (note the "+" and "−" signs next to the relevant synaptic connections, indicating excitatory and inhibitory connections, respectively). This net stimulation sends its activation level *above* 50% by an amount that reflects the degree of the mismatch between the excitatory and the inhibitory signals arriving from the L- and M-cones. That Green/Red opponency cell will then be coding for something in the direction of an increasingly saturated red. Alternatively, if the local preponderance of incoming light favors the medium wavelengths over the long, then the net effect at the Green/Red cell will be *inhibitory*. Its activation level will be pushed below the default level of 50%, and it will then be coding for something in the direction of an increasingly saturated green.

The story for the Blue/Yellow opponency cell is almost identical, except that the rightmost two cone cells (the M- and the L-cones), the ones that are jointly tuned to the *longer* half of the visible spectrum, here join forces to test their joint mettle against the inputs of the S-cone that is tuned to the *shorter* half of the spectrum. A predominance of the latter over the former pushes the Blue/Yellow opponency cell *below* its 50% default level, which codes for an increasingly saturated blue; and a predominance in the other direction pushes it *above* 50%, which (if the inputs from the M- and L-cells are roughly equal) codes for an increasingly saturated yellow.

Finally, the White/Black opponency cell registers the relative preponderance of light at any and all wavelengths arriving at the tiny area of the retina containing the three cone cells at issue, over/under the same kind

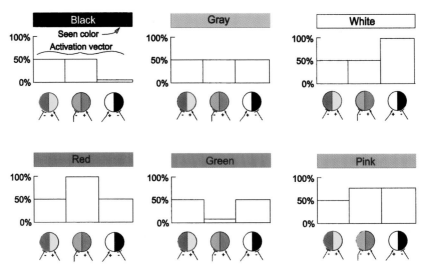

FIGURE 9.3. Vector coding in the H-J network.

of undifferentiated light arriving to the larger retinal area that surrounds the three cones at issue. Accordingly, the White/Black cell performs an assay of the hue-independent *brightness* of the overall light arriving at the tiny retinal area at issue, relative to the brightness of the light arriving to the larger retinal area surrounding it. If that tiny area is much brighter than its surround, the White/Black cell will be pushed above its 50% default level, which codes in the direction of an increasingly bright white. Alternatively, if the tiny area is much *darker* than its comparatively bright surround, then the cone cells within that larger surround will inhibit the White/Black cell below its 50% default level, which codes for an increasingly dark black.

Collectively, the elements of the H-J net are performing a systematic assay of the *power distribution* of the various wavelengths of light incident at its tiny proprietary area on the retina, and an auxiliary assay of the total power of that light relative to the total power of the light incident on its background surround. Distinct objective colors (which implies distinct power distributions within the incident light), when presented as inputs to the net, yield distinct assays at the net's output layer. Six such assays – represented as six histograms – are presented in Figure 9.3, as are the six landmark colors that produce them. To each color, there corresponds a unique assay, and to each assay, there corresponds a unique color.

As you can see, the H-J net converts a four-tuple of inputs – S, M, L, and B (for the level of background illumination) – into a *three*-tuple of outputs: $A_{B/Y}$, $A_{G/R}$, and $A_{W/B}$. Given the several functional relations described in the preceding paragraphs, the activation levels of these three output cells can be expressed as the following arithmetic functions of the four input values:

$$A_{G/R} = 50 + (L - M)/2$$
$$A_{B/Y} = 50 + ((L + M)/4) - (S/2)$$
$$A_{W/B} = 50 + ((L + M + S)/6) - (B/2)$$

These three equations are uniquely determined by the requirement (1) that each of the three second-rung cells has a resting or default activation level of 50%, (2) that the activation levels of every cell in the network range between 0% and 100%, (3) that the different polarities of the several synaptic connections are as indicated in the top part of Figure 9.2, and finally, (4) that each of the three tug-of-wars there portrayed is an even contest.

Very well, but what is the point of such an arrangement? Why convert, in the manner just described, positions in a four-dimensional retinal input space into positions in a three-dimensional opponent-cell output space? The answers start to emerge when we consider the full *range* of possible activation points in the original four-dimensional retinal input space, and the range of their many transformed *daughter* activation points within the three-dimensional opponent-cell output space. To begin, you may observe that those daughter points are all confined within a trapezoid-faced *sub*space of the overall cube-shaped opponent-cell activation space (Figure 9.4a). Points outside that space cannot be activated by *any* combination of retinal inputs, so long as the network is functioning normally, and so long as all inputs consist of reflected ambient light. As written, the three equations cited preclude any activation triplets outside that oddly shaped subspace.

The cut-gem character of that subspace reflects the fact that the three equations that jointly define it are simple linear equations. A somewhat more realistic expression of how $A_{B/Y}$, $A_{G/R}$, and $A_{W/B}$ vary as a function of L, M, S, and B would multiply the entire right-hand side of each of the three equations by a nonlinear, sigmoid-shaped squashing function, to reflect the fact that each of the three output cells is easily nudged above or below its default level of 50%, in the regions close to that level, but is increasingly *resistant* to excursions away from 50% as each approaches the two extremes of 0% and 100% possible activation levels. This wrinkle (for simplicity's sake, I'll suppress the algebra) has the effect of rounding off the sharper corners of the trapezoidal solid, yielding something closer to the spindle-shaped solid with tilted 'equator' portrayed in Figure 9.4b.

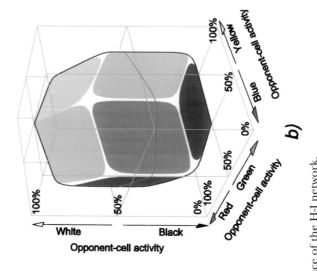

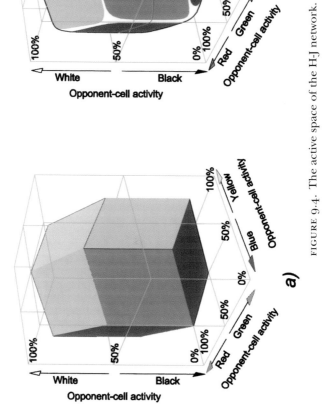

FIGURE 9.4. The active space of the H-J network.

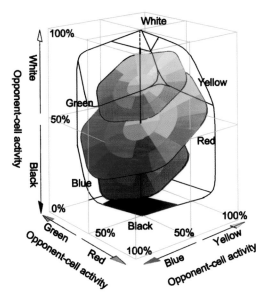

FIGURE 9.5. Color coding in the H-J active space.

Famously, this peculiar configuration of possible *coding vectors* is structurally almost identical to the peculiar configuration, originally and independently reconstructed by Munsell, of possible *color experiences* in normal humans. If one maps the white/black axis of the Munsell solid onto the <50, 50, 100>/<50, 50, 0> vertical axis of the H-J spindle we have just constructed, and the green/red Munsell axis onto the <50, 0, 50>/ <50, 100, 50> horizontal axis of the H-J spindle, and the blue/yellow Munsell axis onto the <0, 50, 35>/<100, 50, 65> tilted axis of the H-J spindle, then the family of distance relations between all of the color experiences internal to the Munsell space is roughly identical with the family of distance relations between all of the coding triplets internal to the H-J spindle.

Note the deliberately color-coded interior of the H-J spindle, which also fades smoothly to white at the top and to black at the bottom, as portrayed in Figure 9.5, and compare it to the interior of Figure 9.1. From precisely such global isomorphisms are speculative thoughts of intertheoretic identities likely to be born. The systematic parallels here described – though highly improbable on purely a priori grounds – become entirely nonmysterious if human color experiences (at a given point in one's visual field) simply *are* the output coding vectors (at a suitable place within some topographical brain-map of the retina) produced by some neuronal instantiation of the H-J net. Such coding vectors presumably reside in a neuronal population fairly early in the human primary visual pathway (e.g., in the parvocelluar ganglion output cells in the retina itself, or in the human LGN, or perhaps in cortical area V_4, three areas where color-opponent cells have been experimentally detected [Zeki 1980], areas whose lesion or destruction is known to produce severe deficits in color discrimination).

This isomorphism of internal relations is joined by an isomorphism in external relations as well. For example, the visual experience of white and the opponent-cell coding vector $<50, 50, 100>$ are both caused by sunlight reflected from such things as snow, chalk, and writing paper. The experience of yellow and the coding vector $<50, 100, 65>$ are both caused by sunlight reflected from such things as ripe bananas, buttercups, and canaries. And so on for the respective responses to *all* of the objective colors of external objects. The a priori probability of these assembled external coincidences is as low as that of the internal coincidences just noted, and their joint (a priori) probability approximates an infinitesimal. These facts most certainly do not entail that the two spaces, and their respective elements, are numerically identical. Other explanations are possible. But we can be forgiven for exploring this most salient possibility.

Two further virtues will complete this brief summary of the H-J net's claim to capture the basics of human color vision. The first additional virtue is the network's capacity for accurately representing the same objective color across a wide range of different levels of ambient illumination. From bright sunlight, to gray overcast, to rain-forest gloom, to a candle-lit room, a humble gray-green object will look plainly and persistently gray-green to a normal human, despite the wide differences in energy levels (across those four conditions of illumination) reaching the three cone-cells on the retina. The H-J net displays this same indifference to variations in ambient brightness levels. Thanks to the tug-of-war arrangement described earlier, the network cares less about the absolute *levels* of cone-cell illumination than it does about the positive and negative *differences* between them.

For example, a cone input pattern of $<L, M, S> = <5, 40, 50>$ will have exactly the same effect at the opponent-cell output layer as a cone input pattern of $<15, 50, 60>$, or $<25, 60, 70>$, or $<43, 78, 88>$, namely, a stable output pattern, at the second layer, of $<A_{B/Y} = 36.25, A_{G/R} = 32.5, A_{W/B} = 50>$, for all four inputs. At least, it will do so if (and only if) the increasing activation levels just cited are the result of a corresponding increase in the level of *general* background illumination. For in that case, the absolute value of B (which codes for background brightness) will also climb, in concert, by 10, 20, and then 38 percentage points as well. This yields incremental increases in inhibition that exactly cancel the incremental increases in stimulation, from the three focal cones, on the White/ Black opponent cell. A color representation that might otherwise have climbed steadily whiteward, into the region of a pastel chartreuse toward the apex of the spindle, thus remains accurately fixed at the true color of the external object – a dull middle green. And this same stability or light-level independence will be displayed for any of the other colors that the network might be called upon to represent.

To cite a final virtue, the H-J net is also roughly stable, in its color representations, across wide variations in the *wavelength profile* of the ambient illumination. At least, it will be thus stable if its constituting cells are assigned the same tendencies to fatigue and to potentiation shown by neurons in general. The human visual system shows the same tendencies, and the same stability. This second form of stability is a little slower to show itself, but it is real. Consider, for example, a nightclub whose ceiling lights emit the lion's share of their illumination at the *long* wavelengths in the visible spectrum. This will provide a false-color roseate tilt to every object in the club, no matter what that object's original and objective color. Upon first entering the club, a normal human will be struck by the nonstandard appearance of every (non-red) object in sight. But after several minutes of adjustment to this nonstandard illumination, the objective colors of objects begin to reassert themselves, the roseate overlay retreats somewhat, and something close to our normal color recognition and discrimination returns, the ever-reddish ceiling lights notwithstanding.

The principal reason for this recovery is that our Green/Red opponent cells, and only those opponent cells, all become differentially *fatigued*, since the nightclub's abnormal ambient illumination forces them all to assume, and to try to maintain, a chronic level of excitation well *above* their normal resting level of 50% – a level of 70%, for example. (That is why nothing looks exactly as it should: every activation triplet produced in this condition, by normal objective colors, will have an abnormal 20% surplus in its $A_{G/R}$ component.) But that 70% level cannot be chronically maintained, because the cells' energy resources are adequate only for relatively brief excursions away from their normal resting level. With their internal resources thus dwindling under protracted demand, the increasingly exhausted Green/Red opponent cells gradually slide *back* toward an activation level of 50%, despite the abnormal ambient light. The artificial 20% surplus in the $A_{G/R}$ component of every coding triplet thus shrinks back toward zero, and the visual system once again begins to represent objective colors with the same coding triplets produced when normal light meets a normal (i.e., unfatigued) visual system. The fatigue accumulated in the system thus compensates, to a large degree, for the nonuniform power distribution across the wavelengths in the ambient illumination.

A symmetric compensation will occur when the ambient light – this time dominated by green, let us suppose – forces the opponent cell to an activation level *below* its normal 50% – to a steady 30%, for example. In that condition, the cell's internal energy resources are not consumed at the normal rate, and they begin to accumulate, within the cell, to abnormal levels. The cell is thus increasingly *potentiated*, rather than fatigued, and its activation levels begin to creep back up toward its default level of 50%, the chronic inhibition of the unusual background light notwithstanding. The same general story holds for the Blue/Yellow opponent cells as well. Taken collectively, these automatic negative and positive compensations allow the H-J net, and us humans, to adapt successfully to all but the most extreme sorts of pathologies in the power spectrum of the ambient light. Such compensations are not perfect, for reasons I shall here pass over.[3] But they are nontrivial. And one need not visit garish nightclubs to have a need for it. In the course of a day, the natural background light of unadorned nature varies substantially in its power spectrum as well as in its absolute power levels. For example, the ambient light under a green forest canopy with a mossy green floor strongly favors the medium wavelengths over the short and the long. But we adjust to it automatically, as described earlier. And the ambient light from the setting Sun favors the longer wavelengths only slightly less than did our roseate nightclub (because the shorter wavelengths are increasingly scattered away as the sinking Sun's light is forced to take a progressively longer path through the intervening atmosphere). But we adjust to this late-afternoon power-spectrum tilt as well.

Colors look roughly normal to us under these two conditions, but they would look arrestingly abnormal without the compensating grace of the adjustments here described. For example, color *photographs* taken in these two conditions will subsequently provide you with an *un*compensated portrayal of the relevant scene (photographic film does

---

[3] Among other things, the input *cones* also become differentially fatigued, but these input cells display a different pattern of compensation. Since their resting activation level is zero, they can display no potentiation, but *only* fatigue.

not fatigue or potentiate in the manner at issue). Such photos will thus be chromatically more hyperbolic – and thus visually more striking – than were your original (fatigue-compensated) visual experiences under the two conditions cited. And if you return to the roseate nightclub for ten minutes or so, there is a further way to appreciate directly just how much your Green/Red opponent cells' current default levels have been pushed away from their normal level of 50%. Simply step outside the nightclub and behold the world in normal daylight. For five or ten seconds, the entire world will have an eerie *greenish* cast to it. This is because every opponent-cell coding triplet, generated in you by light from every seen surface, will have a 20% deficit in its $A_{G/R}$ component, as compared to what an unfatigued system would produce. And such deficits tilt every code toward the code for green.

III. Opponent-Cell Fatigue and Colored Afterimages

This last point will serve to introduce the topic of *colored afterimages*. If a specific area within your visual field (a small circular area, for example) is made deliberately subject to chromatic fatigue or potentiation (by your fixating on a saturated red circle, for example, on a gray background under normal light, for twenty seconds or so), then those differentially fatigued/potentiated opponent cells will yield an appropriately circular afterimage when your gaze is relocated to a uniformly middle-gray background surface. But the apparent color of that afterimage will be the color-complement of the red of the original circular area. That is to say, its apparent color will be at or toward the *antipodes* of the color-spindle position of the red of the original circular stimulus: it will be decidedly green. To illustrate this, look at the first row of Figure 9.6. Fixate for twenty seconds on the small X within the red circle of the leftmost gray square, and then quickly refixate on the X in the middle-gray square two jumps to its right. You will see there a circular green afterimage hovering against that gray background, roughly as portrayed in the final square to the right.[4]

---

[4] The immediate point of placing the colored circle against a middle-gray background square is to ensure that only the visual area comprehending the *circle itself* is subjected to opponent-cell fatigue or potentiation. The immediate point of placing a second, uniformly gray square immediately to the right of the first is to ensure that this square visual area also suffers no opponent-cell potentiation or fatigue. The ultimate point of thus avoiding any fatigue or potentiation in those areas is that, when one's gaze is subsequently refixated on the X within the third square, *everything* within the third and fourth squares will be seen normally *except* the circular area, within the third square, where the induced

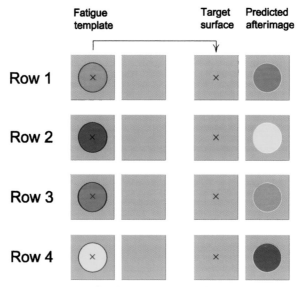

FIGURE 9.6. Elementary color afterimages.

This happens because, when the (now fatigued) opponent cells representing the circular red stimulus are suddenly asked to fall back to representing a less-demanding middle-gray stimulus (as in the third square), they overshoot the required <50, 50, 50> coding vector by an amount equal to whatever fatigue or potentiation has been acquired in each of the three coding dimensions during the protracted exposure to the original red stimulus. That original red stimulus produced an initial coding vector of <50, 95, 50>, but during protracted fixation, that initial vector slowly inches back to something like a vector of <50, 55, 50>, thanks to the accumulated minus-40% fatigue in its middle or $A_{G/R}$ element.

Accordingly, when the opponent cells in the fatigued area are suddenly asked to represent an objectively middle-*gray* stimulus, they can only manage to produce a vector of <50, *10*, 50> – the coding triplet for an obvious middle green – instead of the <50, 50, 50> they would normally produce. For the $A_{G/R}$ cells in the affected circular area are, temporarily, too tired to respond normally. They produce a coding vector with a much-reduced middle component, an abnormal vector that represents green, not gray.

afterimage is situated. The point of the final or right-most square, with its colored circle, is to provide a *prediction* of the shape and expected color of the induced afterimage, next door to it in the third square, so that you may compare directly and simultaneously the reality with the prediction.

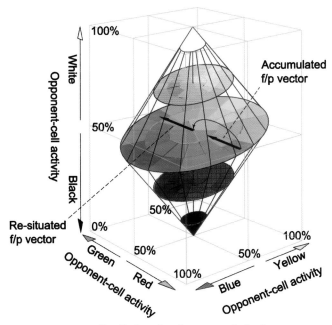

FIGURE 9.7. Predicting the character of afterimages.

(See the fatigue arrows in Figure. 9.7.) Thus, if you fatigue a small part of your visual system by prolonged fixation on a small red circle, you will subsequently see, when you relocate your gaze on a middle-gray surface, a small green circle as an afterimage.

The behavior displayed in this Red/Green example can be generalized to any color whatever, except middle gray itself.[5] Given any position on or toward the outer surface of the Munsell/H-J spindle, a protracted activation triplet starting at that position will slowly creep back toward the middle-gray position at the central core of the spindle. Put another way, for any extremal activation triplet whatever, across the second-rung opponent cells of the H-J network, a *fatigue/potentiation vector* gradually accumulates, a directed line whose arrowhead always points toward the <50, 50, 50> center of the color spindle, and whose tail is always located at the original, extremal activation triplet. When the network

---

[5] Note well that an activation level of 50% of maximum produces neither fatigue nor potentiation in the relevant opponent cell. For under normal conditions, 50% just is the spontaneous resting level of any such cell. Absent any net stimulation or inhibition from the retinal cones, the opponent cells will always return, either immediately or eventually, to a coding vector of <50, 50, 50>, that is, to a middle gray.

is then suddenly given a middle-gray stimulus, the activational result across the opponent cells is always equal to <50, 50, 50> *plus* whatever fatigue/potentiation(f/p) vector $<f_{B/Y}, f_{G/R}, f_{W/B}>$ has accumulated during the protracted fixation on the original extremal color.[6] The abnormal coding triplet (= color-spindle position) that finally results will thus always be directly opposite the original protracted coding triplet, at a distance from the spindle's (gray) center that is equal to the length of the accumulated fatigue/potentiation vector. Return to the final three rows of Figure 9.6, and repeat the experiment for each of blue, green, and yellow circles as the initial fatigue inducer.

A simple rule will convey the point visually (see again, Figure 9.7). For any protracted color stimulus, pick up its accumulated f/p vector as if it were a small arrow pointing rigidly in a constant direction in absolute space, and then place the tail of that arrow at the center of the color spindle, which is the proper coding point for middle gray. The tip of the arrow will then be pointing precisely at the color of the afterimage that will be seen when one's gaze is redirected to a middle-gray surface. Repeat the exercise for a protracted fatigue/potentiation on blue. That f/p arrow will point from the blue periphery of the spindle, toward middle gray. Now pick it up and place its tail on middle gray. That arrow's head will come to rest on yellow, which will be the color of the afterimage that results from an original fixation on blue.

Quite evidently, if a middle-gray surface is the default background against which any colored afterimage is evaluated, then the apparent color of the afterimage must always be located toward a point on the color spindle that is exactly antipodal to the original color stimulus. In any case, we have here one further family of predictions, well known to visual scientists, where the predictive power of the H-J net comes up roses.

---

[6] Note that each element of this f/p vector – $< f_{B/Y}, f_{G/R}, f_{W/B} >$ – can have either a negative value (indicating fatigue for that cell) or a positive value (indicating potentiation for that cell). Note also that the length of that f/p vector will be determined by *(1) how far away* from the spindle's middle-gray center was the original fixation color, and by *(2) how long* the opponent cells were forced to (try to) represent it. A brief fixation on any color close to middle gray will produce an f/p vector of negligible length. By contrast, a protracted fixation on any color far from middle gray will produce an f/p vector that reaches almost halfway across the color spindle. Strictly speaking then, the three equations for $A_{B/Y}$, $A_{G/R}$, and $A_{W/B}$ cited earlier should be amended by adding the appropriate f/p element to the right-hand side of each. This might seem to threaten extremal values below zero or above 100, but the (suppressed) squashing function mentioned earlier will prevent any such excursions. It asymptotes at zero and 100, just as required.

IV. Afterimages Located on Non-gray Backgrounds

We have seen how the f/p vector produced by protracted fixation on any non-gray stimulus will produce the full range of antipodal afterimages when our gaze is subsequently redirected to a neutral (middle-gray) background surface. But there is no reason to limit ourselves to locating and evaluating our afterimages against that background alone. We can locate an acquired f/p vector against a background of any color we like, and we will get a differently colored afterimage for each such differently colored background. The rule established for middle-gray backgrounds holds for any background whatever. Simply add each element of the relevant f/p vector, $< f_{B/Y}, f_{G/R}, f_{W/B} >$, to the corresponding element of the opponent-cell activation vector, $< A_{B/Y}, A_{G/R}, A_{W/B} >$, that would *normally* be produced (in an unfatigued visual system) by the now-colored background surface at issue. That sum will characterize the apparent color of the afterimage as projected against that particular colored background.

For the same reasons, the visual trick cited earlier can also be trusted to give the appropriate predictions here. For any given case, simply pick up the rigidly pointing f/p "arrow" located within the color spindle, and relocate its tail at the color of the chosen background. The arrow's head will then lie exactly at the apparent color of the afterimage that will appear against that chosen background.

Several illustrations of this background-sensitivity in the apparent color of an afterimage are displayed in Figure 9.8. The case of row 1 may surprise you slightly. Fixate at length on the X in the pink circle, and then suddenly refixate on the X in the identically pink square two jumps to the right. You will there find a distinctly *gray* circle hovering over the pink square, much as portrayed in the fourth (predictive) square to its right. This illustrates directly the general principle that acquired f/p vectors always point toward middle gray (see Figure 9.9). As you fixate at length on the original pink circle, you don't realize that your chromatic representation for that area is slowly fading. But you can see instantly that it has indeed faded when it is suddenly relocated against a larger unfatigued background of exactly the same original color. The color *contrast* is obvious: the fatigued area looks grayish.

In row 2, the same f/p vector (again acquired during fixation on the pink circle) is subsequently overlaid on a white background square. The relevant prediction is given by picking up that vector and relocating its tail at the uppermost tip of the color spindle (i.e., at white), as in Figure 9.9.

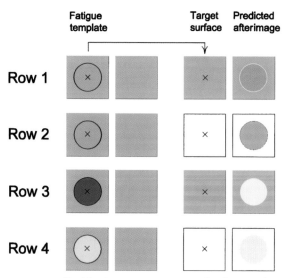

FIGURE 9.8. Afterimages on non-gray backgrounds.

The head of the arrow will then rest at a very pale green. And that will be the apparent color of the circular afterimage.

In row 3 of Figure 9.8, a new f/p vector, generated this time by fixation on a dark blue, is to be relocated on a square background of light blue. Placing the tail of the relevant f/p vector at the light-blue position leaves its arrowhead pointing close to the top of the spindle (i.e., to white), but not quite getting there, as in Figure 9.9. Thus the off-white afterimage that results.

In row 4, a final f/p vector, generated by a bright yellow, is to be relocated on a white background. Moving that vector's tail to the top of the spindle leaves its arrowhead resting at pale blue. And pale blue is the color of the afterimage. (Row 4's vectors – both acquired and resituated – have been left out of Figure 9.9 to avoid clutter.)

Evidently, there are a great many more predictions implicit in the model network at issue. If one considers only a very coarse partitioning of the color spindle (five gray-scale positions on the vertical axis, twelve hue stations around the maximally saturated equator, twelve stations of dullish hue just inside the equator, twelve stations of pastel hue just above it, and twelve stations of darkish hue just below it), one is looking at a total of 53 distinct colors on which to fixate at length, and 53 possible colors on which to locate the resulting afterimage. The possible combinations total 53^2, or 2,809 distinct experiments, each one a test of the H-J hypothesis

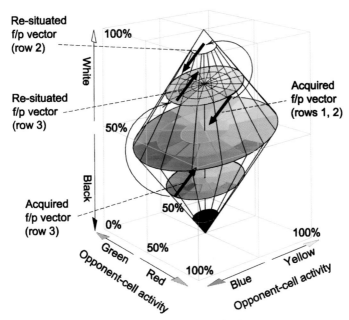

Re-situated
f/p vector
(row 2)

100%

White

Re-situated
f/p vector
(row 3)

50%

Black

Acquired
f/p vector
(row 3)

50%

0%

Acquired
f/p vector
(rows 1, 2)

100%

Green
Opponent-cell activity

Red

50%

100%

Yellow

Blue
Opponent-cell activity

FIGURE 9.9. Nonantipodal afterimages.

about human color coding. I have personally tested 280 (or 10%) of them. The H-J net's predictive performance is both systematic and strikingly accurate. But there is more to come.[7]

V. Breaking Out of the Color Spindle: Chimerical Colors

It was remarked in Section II that the equations governing the H-J net guarantee that any activation triplet within the opponent-cell activation

---

[7] The wary reader may have noticed that I am assuming that it is *opponent-cell* fatigue/ potentiation, as opposed to *retinal cone-cell* fatigue, that is primarily responsible for the chromatic appearance of our afterimages. Why? For three reasons. First, we still get strongly colored afterimages even at modest light levels, under which condition the cone cells are *not* put under stress, but the delta-sensitive opponent cells regularly are. Second, as I noted earlier, the opponent cells code by variations in high-frequency spiking, which is much more consumptive of energy than is the graded-voltage coding scheme used in the cones. Accordingly, the opponent cells are simply more *subject* to fatigue/potentiation. Finally, the H-J theory entails one pattern of afterimage coloration if cone-cell fatigue is the primary determinant, and a very different pattern of afterimage coloration if opponent-cell fatigue/potentiation is the primary determinant. The observed pattern of coloration agrees much more closely with the latter assumption. Colored afterimages, it would seem, are primarily an opponent-cell phenomenon.

space would be strictly confined to the subspace that constitutes the classical color spindle, no matter what combination of cone-cell activities produced that triplet. As those equations are written, that observation is correct, and it serves to explain the gross shape of Munsell's original spindle, including its tilted equator that makes saturated blue a much darker color than saturated yellow.[8] But you may still want to ask, what about all that *unused* space in the several upper and lower corners of the opponent-cell activation cube? What would be the significance of a possible activation triplet *outside* the classical color spindle, a triplet somewhere in that fairly considerable volume of unused opponent-cell activation space?

Well you might ask. In particular, you might ask after the *phenomenological* significance of such an extraspindle activation vector. Would it still be a color appearance of some sort, but chromatically distinct from anything within the spindle? Would it still follow or respect the basic rules of color similarities and differences that hold within the spindle? What would it 'be like' to have such an activation vector realized in one's own opponent cells? If the H-J account of things is even roughly correct, you are about to find out.

Inserting stimulating/inhibiting electrodes directly into some substantial population of opponent-cell neurons in a human would afford us the opportunity to produce directly, and independently of the peculiar connectivity of the H-J net, any activation vector that we choose. In principle, it could be done. But capturing a sufficiently large *number* of cells simultaneously (after all, the anomalous chromatic area within one's subjective visual field has to be large enough for one to discriminate) is currently beyond our experimental technologies, and the cranial invasion would needlessly threaten the health of the human subject in any case.

Fortunately, there is a much easier way to produce the desired result. In the last several figures, you have already been introduced to the required

---

[8] The explanation is obvious. Recall that to produce an opponent-cell triplet for maximum white requires that *all three* of the S-, M-, and L-cones have activation levels of 100. To produce a triplet for maximum black requires those same cells all to be at zero. To produce a triplet for saturated yellow requires that S be at zero, while M and L are *both* at 100. Accordingly, the retinal input for yellow already takes you two-thirds of the way up the opponent-cell cube toward white (recall that white requires S, M, and L all to be at 100). Similarly, the input for saturated blue requires that S be at 100, while both M and L are at zero. Accordingly, the retinal input for blue already places you two-thirds of the way toward black (which requires that S, M, and L are all at zero). Hence, blue is darker than yellow and the equator of maximum hue saturation must be tilted so as to include them both. (Note that red and green display no such brightness asymmetry.)

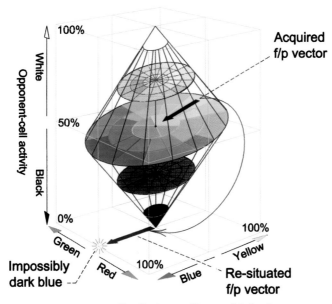

FIGURE 9.10. Predicting an "impossible" color.

technology, namely, selective fatigue/potentiation by prolonged fixation on some suitable color stimulus. Recall that the opponent-cell activation vector that would *normally* result from a given retinal stimulus is subject to substantial *modification* by the addition of whatever f/p vector has accumulated in the system immediately prior to that external retinal stimulus. Adding those two vectors together can indeed yield a vector that reaches well outside the classical spindle. For example, let the system fatigue on yellow, as indicated in Figure. 9.10. Then present the system with a maximally *black* stimulus. The resulting vector will reach out from the bottom tip of the spindle, along the floor of the opponent-cell activation space, to a place directly but distantly *underneath* the standard coding triplet for a maximally saturated blue, as also indicated in Figure 9.10.

Extrapolating from what we already know about the coding significance of the three major dimensions of the color spindle and of the H-J opponent-cell activation space, that anomalous activation triplet must code for a color appearance that is

1. fully as dark as the darkest possible black (it is, after all, on the maximally dark *floor* of the opponent-cell activation space), but nevertheless is of

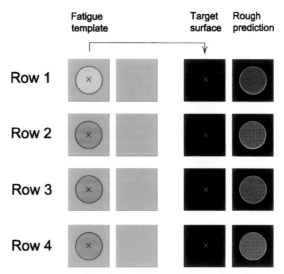

FIGURE 9.11. Producing "impossibly dark" colors.

2. an obvious and distinctive hue (it is, after all, on a radius quite far from the hue-less central axis of the opponent-cell activation space), a hue that must be

3. more similar to blue than to any other hue around the spindle's equator (it is, after all, closer to blue than to the position of any other such color).

On the face of it, the joint satisfaction of these descriptive conditions might seem to be impossible, for no *objective* hue can be as dark as the darkest possible black, and yet fail to *be* black. As the original Munsell color spindle attests, to get to anything that has an objective hue, one must leave the hue-less central axis in some horizontal direction or other. But to do that, one must come up that brightness axis, at least some distance, if one is to escape the bottommost singularity of maximal black.

However, we are not talking about the objective colors of real objects at this point. We are talking about an anomalous *color representation* within the cubical opponent-cell activation space. And these anomalous representations are robustly possible, as you can discover for yourself in row 1 of Figure 9.11. A twenty-second fixation on the X at the center of the yellow circle will produce in you precisely the f/p vector at issue. And your subsequent fixation on the X at the center of the maximally black square to its right will produce in you precisely the anomalous, extraspindle coding vector here under discussion. (It fades away after a few seconds, of course, as the relevant cells progressively recover from their induced fatigue/potentiation.)

The final black square to the far right of row 1 contains, as before, a (very rough) prediction of what your circular afterimage will look like. But here my prediction-image is doomed to be inaccurate, for the very dark blue circle there inscribed is still objectively and detectably brighter than its black surround. (It could not be otherwise without losing its blue hue entirely.) The anomalous afterimage, by contrast, presents a circular patch that is every bit as dark as its black surround, and yet appears decidedly blueish in some unfamiliar way. It is also visibly *darker* than the dark-blue 'predictive' circle to its immediate right. That afterimage meets, while the (roughly) predictive objective image does not and cannot meet, all three of the conditions (1)–(3) listed earlier. This provides you with an experience of what might be called a "chimerical color" – a color that you will absolutely never encounter as an objective feature of a real physical object, but whose qualitative character you can nonetheless savor in an unusually produced illusory experience.

(You may have to resist an initial temptation to judge the anomalous afterimage to be at least somewhat brighter than its black surround *on the grounds that* anything with a detectable hue *must* be somewhat brighter than black. But the principle that would normally license that inference is valid only for objective colors, not for internal color representations. Repeated examination of the circular afterimage will reveal that it is indeed as dark as its maximally dark surround, despite its vivid saturation, and that it is always much darker than the [inadequate] dark-hue prediction-circle to its immediate right. In sum, these weird afterimages are definitely outside the classical spindle. They are not just pressing at its periphery.)

Rows 2–4 of Figure 9.11 provide the resources for three more 'chimerically colored' afterimages. The second yields an impossibly dark but still-vivid green. The third yields an impossibly dark but still somehow-vivid red. And the fourth yields, with maximum implausibility, a yellow that is as dark as the darkest possible black, and yet is still not black. You may well judge it to be some kind of unfamiliar *brown*, rather than yellow. (In general, the visual system sees things as brown exactly when it is confronted with an external stimulus that returns the same *wavelength profile* as yellow, or orange, but which is also judged, by contrast effects, to have a very low overall intrinsic reflectance. This fits the case at hand.)

The theory predicts, of course, the existence of impossibly dark versions of all of the hues around the classical spindle's equator, not just the

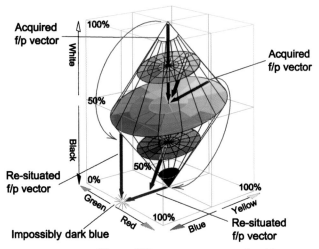

Acquired f/p vector

100%

Acquired f/p vector

White

50%

Black

Re-situated f/p vector

0%

50%

100%

Green

Red

100%

Blue

Yellow

100%

Re-situated f/p vector

Impossibly dark blue

FIGURE 9.12. Three different routes to stygian blue.

canonical four we have examined here. Fixate at length on any saturated hue of your choosing; locate the resulting afterimage over a maximally black background, and the 'seen color' will be an impossibly dark version of the color-complement of the original circular color stimulus.

The theory also predicts that there exists *more than one way* to produce a given chimerical color sensation. Figure 9.12 illustrates the point. One can produce a chimerical dark blue by fatiguing on a saturated yellow stimulus, and then locating the afterimage over a black background (as we have already seen in Figure 9.11, row 1). But one can produce the same result by fatiguing on a bright *white* stimulus, and then locating the afterimage over a saturated *blue* background. You get to the same place, but by a very different route. Indeed, one can produce (almost) the same result by fatiguing on a pastel yellow stimulus, and then locating the afterimage on a *dark gray* background.

You may test all three of these predictions simultaneously by fixating at length on the pastel yellow stimulus in row 2 of Figure 9.13. This will produce in you three afterimages at once, arranged in a vertical line. When you then locate the middle row's afterimage over the dark-gray background to the right, the first and third row's afterimages will be located over the blue and the black backgrounds, respectively. You can then compare the qualitative character of all three afterimages at once, at least for the few seconds before they begin to fade.

The theory predicts that the afterimages in rows 1 and 3 will be the darkest, and the most strikingly blue. The afterimage in the middle should

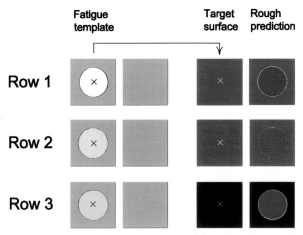

FIGURE 9.13. Producing three stygian blues simultaneously.

be less dramatic, both in its darkness and in its blue hue, for the pastel yellow stimulus that originally produced it was not quite adequate to produce the maximal fatigue achieved by the other two stimuli. The theory also entails a different pattern of *fading* for each of the three afterimages. The afterimage in row 1 will slowly fade in its degree of darkness, but not in its degree of blueness. (Its fatigue lies in the black/white dimension.) By contrast, the afterimage in row 3 will progressively fade in its blue hue, but not in its degree of darkness. (Its fatigue lies in the blue/yellow dimension.) Finally, the afterimage in row 2 will progressively fade in both its hue *and* its darkness. (Its fatigue lies in both of those dimensions.) You may repeat these exercises, of course, with any other hue around the spindle's equator.

Such fine-grained predictive prowess, concerning both these unusual qualitative characters and the various changes they display over time, is noteworthy. But there is still more to come.

VI. Out of the Spindle Again: Self-Luminous Colors

Let us not forget the *upper* regions of the opponent-cell activation space (see Figure 9.14). Prolonged fixation on a red circle will produce an f/p vector that, when its tail is relocated to the upper tip of the spindle, will reach out horizontally, across the ceiling of the space, to a point that represents a color that is as bright as the brightest white (after all, it is on the ceiling). But it cannot be white (after all, it is some distance away from the hue-less central axis). Instead, it must be some implausibly luminous

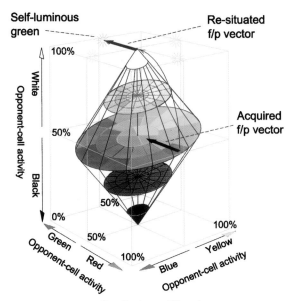

FIGURE 9.14. Predicting self-luminous green.

cousin of green. See for yourself in row 1 of Figure 9.15. As before, fixate on the central X for at least twenty seconds, and then refixate on the central X of the white target square to its immediate right. Here you will notice that the bright-green(ish) afterimage seems positively *self-luminous*,

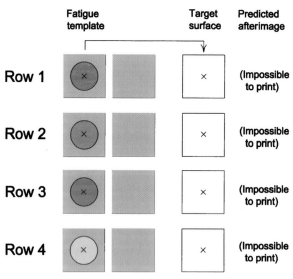

FIGURE 9.15. Producing "self-luminous" colors.

as if it were a colored lightbulb or a colored LED (light-emitting diode). The impression of faint self-luminosity here is entirely understandable, for no physical object with a detectable objective hue could be possibly be as bright as a maximally white surface unless it were in fact *self*-luminous, *emitting* even more light energy than a white surface could possibly reflect under the ambient lighting.

Row 2 displays the same phenomenon, but with a blue fatigue template. In this case the afterimage is an apparently self-luminous yellow. Other instances of apparently self-luminous afterimages can be produced by prolonged fixation on any hue whatever, so long as the hue's $A_{W/B}$-component is at or below 50%. (Otherwise the acquired f/p vector will have a nonzero downward component that will force the resulting afterimage down and away from the maximally bright ceiling of the opponent-cell activation space. The illusion of self-luminosity will progressively fade.) Rows 3 and 4 will complete the chromatic quartet of apparently self-luminous afterimages.

Once again, we are contemplating color qualia whose location in qualia space (= opponent-cell activation space) lies well outside the classical color spindle. Their existence is not quite the curiosity that their impossibly dark basement-floor cousins were, but that is because we have all encountered them in common experience. Munsell's original concerns were confined to the colors of *non*-self-luminous Lambertian (light-scattering) surfaces. Chromatic phenomena may begin there, but they do not end there. Self-luminous colors occur when an object emits (rather than merely reflects) a nonuniform wavelength profile at an energy level that is too high (i.e., too bright) to be accounted for by the maximum reflection-levels possible given the ambient or background illumination (i.e., by a white surface). Such unusual stimuli must therefore be coded at the absolute ceiling of the opponent-cell activation space, but (because of the nonuniform wavelength distribution) it must be coded at some appropriate distance away from the hue-neutral center of that ceiling.

A normal Lambertian surface meeting a normal visual system will never produce such a coding triplet, no matter what the ambient illumination. But a self-luminous colored object will certainly do so, even in a normal visual system, for the stimulation-levels it produces in the cones exceeds anything that the ambient or background light-levels could possibly produce. Such an extraspindle coding triplet is thus a signature sign of self-luminance in the environment. And that is why, when we produce such ceiling-dwelling coding triplets artificially, as in the fatigue/potentiation experiments of this section, the immediate impression is of a self-luminous colored object.

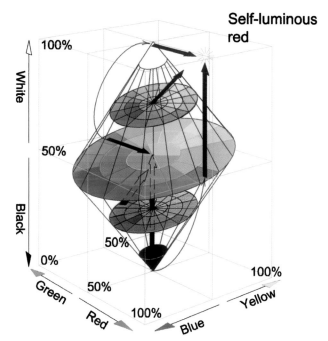

FIGURE 9.16. Three ways to "self-luminous" red.

Note also that one can produce a sensation of the same anomalously bright color in more than one way, as illustrated in Figure 9.16. Fixate on a green circle, for example, and then look at a white surface, as in row 1 of Figure 9.17. Alternatively, fixate on a black circle, and then look at a red surface, as in row 3. Both procedures produce the same afterimage. Evidently, the family of predictions explored near the floor of the opponent-cell activation space is reflected in a similar family of predictions concerning the behavior of sensations at its ceiling. These, too, test out nicely.

VII. Out of the Spindle One Last Time: Hyperbolic Colors

Return your attention to the central plane of the color spindle. Note that the coding triplets for the maximally saturated versions of the four primary hues – green, red, blue, and yellow – are all hard-pressed against the outer walls of the all-inclusive opponent-cell activation cube. By contrast, the four intermediate hues – yellow-green, orange, purple, and blue-green – all look out on some normally unused space in the four corners of the central horizontal plane. Using the techniques already explored, we

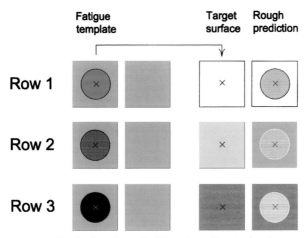

FIGURE 9.17. Producing three "self-luminous" reds simultaneously.

can contrive to activate a coding triplet in any one of those four extremal corners. If one fixates at length on a pale blue-green stimulus, as in Figure 9.18, and then refixates on an *already* maximally saturated orange surface, then the resituated f/p vector will yield an activation triplet within the farthest corner of the cube, beyond the limits of the classical spindle. The H-J theory of our internal color representations entails that one should there find a circular afterimage of a *hyperbolic* orange, an orange that is more 'ostentatiously orange' than any (non-self-luminous) orange you have ever seen, or ever will see, as the objective color of a physical object. Row 1 of Figure 9.19 will allow you, once more, to test such a prediction for yourself.

Row 2 provides access to a similarly hyperbolic version of purple. Rows 3 and 4 jointly provide a *non*hyperbolic contrast to the first two rows. Here we are set up to try to produce a hyperbolic red and a hyperbolic green, respectively. But here the theory says that there is little or no room inside the opponent-cell activation cube, beyond saturated red and saturated green, where any such hyperbolic activation triplet might locate itself. The coding triplets for saturated red and saturated green are *already* at, or close to, the relevantly extremal positions. And so here the theory predicts that our attempts to find a hyperbolic red and green must fail – or, at least, find a much feebler success than we found in the cube's more capacious corners. See what you think about the relative vividness of the afterimages in the last two rows, relative to those achieved in the first two rows.

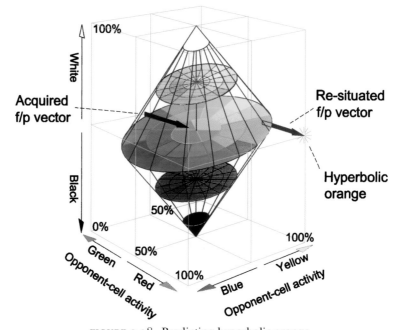

FIGURE 9.18. Predicting hyperbolic orange.

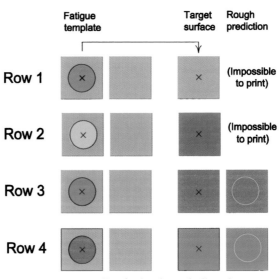

FIGURE 9.19. Producing hyperbolic colors.

VIII. The Consequences for Current Philosophical Debates

The reader will note that, despite the nontrivial (but wholly defeasible) case laid out earlier, in support of the strict identity of human visual color qualia on the one hand and human opponent-cell coding triplets on the other, at no point did we establish, or even try to establish, that there is any sort of *necessary connection* between the two. I did not argue, nor claim, that the former are 'logically supervenient' upon the latter (cf. Chalmers 1996). I did not argue, nor do I believe, that the identity at issue is blessed by any form of 'metaphysical necessity' (cf. Kripke 1972). Nor did I suggest that there is any form of 'lawlike' or 'nomological' connection between the two (cf. Davidson 1970). As I have argued elsewhere, all of these diverse modal relations are philosophical extravagances or confusions imposed, post facto, on successful cases of historical intertheoretic reductions, all of which were achieved without the help of such modal relations, and *none of which displays any one of them* (cf. Churchland 1979, 1985, 1996c). Here, as in those other cases from our scientific history, the principal intellectual motive for embracing the systematic color-qualia/coding-vector identities proposed is simply the extent and quality of the predictive and explanatory unity that the relevant reduction provides.

But that basic motive was already in place, independently of the experimental predictions of the present paper. If those predictions are correct, they provide an *additional* motive for embracing the proposed reduction of color qualia to coding vectors. For it was no part of the motives – for the H-J net's original reductive proposal – that these particular experimental predictions be a part of the explanatory target. They were unanticipated, and they are faintly paradoxical on their face. They thus provide some "excess empirical content" beyond the original explanatory target, namely, our familiar experiences of the mundane colors of external objects.

Such excess empirical contents are familiar from the history of science. In the latter part of the nineteenth century, the assumption that light was identical with electromagnetic waves entailed that there should be such a thing as *invisible* light (an apparent contradiction, note well). Specifically, there should be light with a wavelength longer than the red end (namely, infrared light), and there should be light with a wavelength shorter than the violet end (namely, ultraviolet light), of the visible spectrum (.40μm to .70μm). Despite this clear violation of then-normal semantic expectations, the existence of light – not faint light, but very bright light – *outside* the visible spectrum was subsequently confirmed by Herschel, Hertz, and Roentgen. The elusive and apparently singular nature of light was thus brought under the broad umbrella of electromagnetic phenomena in general. Not to its detriment, but to its welcome illumination.[9]

The parallel assumption, that human color-representations or color-qualia are identical with opponent-cell coding triplets in a neuronal instantiation of the H-J network, yields a similarly implausible prediction. There should exist color-qualia outside the qualitative range of the classical color spindle, qualia whose perfectly accurate descriptions violate our normal semantic expectations. The H-J theory further suggests how to produce such chimerical qualia – through opponent-cell fatigue/potentiation – so that we may test those unexpected predictions against our own experience.

Let me now point out that, just as in the case of light, the new theory also provides a wealth of *explanatory* power commensurate with its extensive predictive power. Why, in the case of Figure 9.11, row 1, is one's stygian circular afterimage an image of something so similar to objective *blue*? Because the coding vectors for the opponent cells in that part of your visual field all have an $A_{G/R}$ component that is neutrally balanced at 50%, and an $A_{B/Y}$ component that is well *below* 50%. Such vectors are precisely those that code for the various blues.

Why is that $A_{B/Y}$ component so unusually low? Because the relevant cells were antecedently fatigued on a maximally yellow stimulus, which forced them to try to maintain an $A_{B/Y}$ component close to 100%. Their subsequent response in that dimension, to any external stimulus, will thus be much lower than normal, at least for a short time.

Why is the seen blue so curiously and implausibly *dark*? Because in this case the $A_{W/B}$ component of the relevant activation vector is close to 0%, which makes it similar, in its darkness, to a sensation of maximal *black*.

Why does that impossibly dark blue afterimage fade over five seconds or so, as I gaze at the black square target? Because the B/Y opponent cells (no longer under input pressure to maintain an extreme value) slowly recover from their fatigue, and slowly return to representing the black background for what it really is: a maximal black.

---

[9] Herschel placed the bulb of a mercury thermometer just outside the redmost edge of the spectral 'rainbow' image produced by directing sunlight through a prism. The mercury level shot up. Hertz confirmed the existence of light at much longer wavelengths with his primitive radio transmitter and radio receiver. Roentgen stumbled across x-rays while playing with a cathode-ray tube, and he correctly characterized them, after a week or two of sleuthing, as light of much shorter than visible wavelengths.

Why does the initial saturation level of the 'impossibly' blue afterimage depend on *how long* I stared at the yellow circle? Because the degree of fatigue induced in the B/Y cells is a function of how long they were forced (by a maximally yellow stimulus) to try to maintain an unusually high level of activation. The greater the fatigue, the more abnormally *low* will be their subsequent response to any stimulus. And the farther from the 50% neutral point they fall, the greater is the saturation of the stygian blue therein represented.

Evidently I could go on illustrating the H-J net's explanatory virtues – concerning the qualitative characters and qualitative behaviors of *thousands* of colored afterimages – but you can now see how to deploy the explanatory virtues of that model for yourself. The point of the preceding paragraph is to underscore the claim that the theory here deployed has just as much explanatory power as it has predictive power. The alleged 'explanatory gap'[10] intruding between physical theory and phenomenological reality turns out to be a reflection of nothing more than our own failures of explanatory imagination and an inadequate understanding of the human nervous system.

A caution: as colorful as they might be, one's individual reactions to the several tests set out in this paper (Figures. 9.6, 9.8, 9.11, 9.13, 9.15, 9.17, and 9.19) should not be regarded as adequate grounds for believing the theory here deployed. The H-J theory has an independent authority, derived from many prior tests over the past two decades, as the first half of this paper attempts to summarize. And the issue of chimerically colored afterimages in particular wants attention from competent visual psychologists who can bring the appropriate experimental procedures to bear on a sufficiently large population of naïve subjects. But you may appreciate, from the simple tests here provided, why such experimental work might be worth doing. And you may also appreciate my own expectations in this matter. Indeed, by now you may share them.

Withal, how those experiments actually turn out is strictly beside the philosophical issue that opened this paper. We began by confronting the philosophical claim that no physical theory could ever yield specific predictions concerning the qualitative nature of our subjective experience. And yet here we have before us a theoretical initiative that yields precisely the sorts of predictions – in qualitative detail – that were supposed to be impossible. Moreover, the predictions at issue concern genuinely

---

[10] For the *locus classicus* of this worry, see J. Levine, "Materialism and Qualia: The Explanatory Gap," *Pacific Philosophical Quarterly* 64 (1983): 354–61.

novel phenomena, phenomena beyond our normal qualitative experience. And at first blush, it seems that those predictions might even be true.

That would indeed be interesting. But it is not strictly the point. The point is that a sufficiently fertile theory of chromatic information processing in the human visual pathway (i.e., the H-J model) coupled with a sufficiently systematic grasp of the structure of the explanatory target domain (i.e., the Munsellian structure of our phenomenological quality space) can, when fully explored, yield predictions that could never have been anticipated beforehand, predictions that would have been summarily dismissed as "semantically odd" or outright impossible, even if they had been anticipated.

This lesson is as true, and as salutary, in the present and rather more modest case of subjective color qualia as it was in the nineteenth-century case of light. Apparent 'explanatory gaps' are with us always and everywhere. Since we are not omniscient, we should positively expect them. And here, as elsewhere, apparent (repeat: *apparent*) qualitative 'simples' present an especially obvious challenge to our feeble imaginations.[11] But whether an apparent gap represents a mere gap in our current understanding and imaginative powers, or an objective gap in the ontological structure of reality, is always and ever an empirical question – to be decided by unfolding science, and not by preemptive and dubious arguments a priori. In light of the H-J network's unexpectedly splendid predictive and explanatory performance across (indeed, *beyond*) the entire range of possible colors, the default presumption of some special, nonphysical ontological status for our subjective color experiences has just evaporated. Our subjective color experiences – the chimerical ones, included – are just one more subtle dimension of the labyrinthine material world. They are activation vectors across three kinds of opponency-driven neurons. This should occasion neither horror nor despair. For while we now know these phenomenological roses by new and more illuminating names, they present as sweetly as ever. Perhaps even more sweetly, for we now appreciate why they behave as they do.

I conclude by addressing a final objection to the specific identities here proposed. "We can see why you propose to identify the subjective qualia of saturated redness with an opponent-cell activation vector of $<50, 100, 50>$ (as in Figure 9.3), and so on for all of the other elements

---

[11] On this point in particular, see P. M. Churchland, "The Rediscovery of Light," *Journal of Philosophy* 93, no. 5 (1996), sec. V, "Some Diagnostic Remarks on Qualia," 225–8.

of the proposed mapping. It is because this mapping has the virtue that all of the proximity (similarity) relations within qualia space are successfully mirrored in the assembled proximity (similarity) relations within the relevant activation-vector space. But a problem remains. To begin, there is no guarantee that this particular mapping is the *only* mapping that would achieve that end. Perhaps there are others, as contemplated in the familiar class of 'inverted spectrum' thought experiments (cf. Chalmers 1996). More specifically, the account proposed in the preceding pages fails to give an adequate explanation, or indeed, *any* explanation, of why an activation vector of <50, 100, 50> should have, or produce, or be associated with, a qualitative character of *this* particular nature (I here inwardly advert to what I have learned to call 'a sensation of *red*'), as opposed to any of the other available color qualia. In the absence of such an explanation, the account of the preceding pages has uncovered nothing more than a systematic but still-puzzling empirical *correlation* between qualia on the one hand, and opponent-cell activation vectors on the other. What qualia might be, in themselves, remains a mystery."

This deflationary complaint is seductive, but it betrays a fundamental misunderstanding of what is going on in any proposed intertheoretic reduction, and of the requirements that any reduction must meet in order to be successful. The demand for an *explanation*, as outlined in the preceding paragraph, is ill-conceived for precisely the case of the intertheoretic identities at issue. This is not hard to see. To ask for an explanation of why a given qualia is 'correlated' with a given activation vector is to ask for some natural law or laws that somehow *connect* qualia of that kind with activation vectors of the relevant kind. But there can be such a natural law only if the quale and the vector are *distinct things*, things fit for enjoying nomic connections with one another. In the case at issue, however, the proposal is that the qualia and the vectors are not distinct things at all: they are identical; they are one and the same thing, although known to us by two different names. An explanation of the kind demanded is thus impossible, and the demand that we provide it is misconceived from the outset. As well demand a substantive explanation for the curious and universal co-occurrence of the substance *snow* and the substance *neige*.

Granted, the respective background conceptual frameworks that embed the notions of qualia, on the one hand, and activation vectors, on the other, are much more different from one another than are the respective English and French conceptions of snow. But that is precisely why the identifications proposed in the present paper, and those proposed in intertheoretic reductions generally (recall "light = electromagnetic waves," "temperature = mean molecular kinetic energy," and "pitch = oscillatory frequency"), are so much more *informative* than are the identifications made in the humdrum case of closely synonymous translations. They bring new explanatory resources to bear on an old and familiar domain, and they provide novel empirical predictions unanticipated from within the old framework, as this essay has just illustrated. To demand a substantive explanation of the 'correlations' at issue is just to beg the question against the strict identities proposed. And to find any dark significance in the 'absence' of such an explanation is to have missed the point of our explicitly *reductive* undertaking.

Nor need the specter of various possible qualia inversions across distinct individuals, or within a given individual over time, trouble the reductive account here proposed. For the H-J account of subjective color experiences not only allows for the possibility of such inversions *it specifies exactly how to produce them.* For example, if you wish to produce a global green/red inversion in your subjective qualitative responses to the external world (while holding the black/white and blue/yellow dimensions unchanged), simply change the polarity of all of the L-cone projections (to the green/red opponent cells) from excitatory to inhibitory, and change the polarity of all of the M-cone projections (to the green/red opponent cells) from inhibitory to excitatory, and change nothing else, especially in the rest of your visual system downstream from your now slightly rewired opponent cells (see again Figure 9.2). That will do it. Upon waking from this (strictly fanciful) microsurgery, everything that used to look red will now look green, and vice versa.

But there is no metaphysical significance in this empirical possibility, nor in the many other possible inversions and gerrymanderings that similar rewirings would produce. For we are here producing systematic *activation-vector inversions* relative to the behavior of activation vectors in a normal (i.e., un-rewired) H-J network. Given the strict identities proposed between specific qualia and specific activation vectors, it is no surprise that changes in the response profile of either one will be strictly 'tracked' by changes in the other. Of course, it remains an a priori possibility that our color qualia might vary *independently* of the physical realities of the H-J network. But this is just another way of expressing the permanent a priori possibility that the Hurvich-Jameson account of our color experiences might be factually mistaken, and this is something to which everyone must agree. But do not confuse this merely a priori issue with a closely related empirical issue. If the H-J account of our color experiences is correct, then it is empirically *im*possible to change the profile of our subjective color responses to the world without changing in some way the response profile of our opponent-cell activation vectors, as outlined, for example, in the preceding paragraph. From this reductive perspective, sundry "qualia inversions" are indeed possible, but not without the appropriate rewirings within the entirely physical H-J net that embodies and sustains all of our color experience. If we wish to resist this deliberately reductive account – as some still may – then let us endeavor to find in it some real *empirical* failing. Imaginary failings simply don't matter.

10

On the Reality (and Diversity) of Objective Colors

How Color-Qualia Space Is a Map
of Reflectance-Profile Space

Abstract: How, if at all, does the internal structure of human phenomenological color space map onto the internal structure of objective reflectance-profile space, in such a fashion as to provide a useful and accurate representation of that objective feature space? A prominent argument (due to Hardin, among others) proposes to eliminate colors as real, objective properties of objects, on grounds that nothing in the external world (and especially not surface-reflectance profiles) answers to the well-known and quite determinate internal structure of human phenomenological color space. The present paper proposes a novel way to construe the objective space of possible reflectance profiles so that (1) its internal structure becomes evident, and (2) that structure's homomorphism with the internal structure of human phenomenological color space becomes obvious. The path is thus reopened to salvage the objective reality of colors, in the same way that we preserved the objective reality of such features as temperature, pitch, and sourness – by identifying them with some objective feature recognized in modern physical theory.

I. Introduction to the Problem

At least since Locke,[1] color scientists and philosophers have been inclined to deny any objective reality to the familiar ontology of perceivable colors,

---

[1] *An Essay Concerning Human Understanding*, Book II, ch. viii. For the analytic and exegetical case that Locke was indeed an eliminativist, rather than some sort of reductionist, about objective colors, see the thoughtful essay by Samuel C. Rickless, "Locke on Primary and Secondary Qualities," *Pacific Philosophical Quarterly* 78 (1997): 297–319. To be sure, Locke's text admits of other interpretations.

The central idea of this paper occurred to me while I was listening to a provocative talk on color given by Mohan Matthen during the Vancouver Conference on the Philosophy of Color, in October 2003. My thanks to him for his inspiration. The paper also reflects what I have learned over the years about color from Larry Hardin, Kathleen Akins, and Martin Hahn. My thanks to them also.

on grounds that physical science has revealed to us that material objects have no qualitative features at their surfaces that genuinely *resemble* the qualitative features of our subjective color experiences. Objective colors are therefore dismissed as being, at most, "a power in an object to produce *in us* an experience with a certain qualitative character." Accordingly, colors proper are often demoted from being 'primary properties' (i.e., objective properties of external physical objects) to the lesser status of being merely 'secondary properties' (i.e., properties of our subjective experiences only).

To be sure, we are not logically forced to this eliminative conclusion by the failure of the first-order resemblances cited. A possible alternative is simply to *identify* each of the familiar external, commonsense colors with whatever "power within external objects" it is that tends to produce the relevant internal sensation. More specifically, we might try to identify each external color with a specific *electromagnetic reflectance profile* had by any object that displays that color. The objective reality of colors would then emerge as being no more problematic than is the objective reality of the *temperature* of an object (which is identical to the mean kinetic energy of its molecules), or of the *pitch* of a sound (which is identical to the dominant oscillatory frequency of an atmospheric compression wave), or of the *sourness* of a spoonful of lemon juice (which is identical with the relative concentration of hydrogen ions in that liquid). These parallel properties *also* fail the 'first-order resemblance' test imposed by Locke and other early modern thinkers. Nonetheless, their successful reduction to objective properties of material objects is an accomplished fact, both of science and of settled history. Locke's criterion for objective reality – a first-order resemblance to the qualities of our sensations – was simply ill-conceived.

On the more modern reductive approach displayed in these examples, color may turn out to be, by the standards of uninformed common sense, a somewhat surprising sort of feature, namely, a profile of reflectance efficiencies across the visible part of the electromagnetic (EM) spectrum. But this is no more surprising than any of the other identities just cited. And no more surprising, perhaps, than is the identification of light itself with electromagnetic waves. Such identities may surprise the scientifically uninformed, but they leave the objective reality of light, temperature, pitch, and sourness entirely intact.

Unfortunately, this happy (reductive) accommodation would seem to be denied us in the case of colors in particular. For, it is often argued, there *is no* unique EM reflectance profile that corresponds to, and might thus be a candidate for identification with, each (or indeed, any) of the familiar colors. On the contrary, to each of the familiar colors there corresponds an apparently unprincipled variety of decidedly *different* reflectance profiles. The scattered class of such diverse profiles, for each 'objective' color, is called the class of *metamers* for that color, and they are indeed diverse, as the four profiles in Figure 10.1 illustrate.

Four distinct material objects, each boasting one of the four reflectance profiles here portrayed, will appear identically and indistinguishably yellow to a normal human observer under normal illumination (e.g., in broad daylight). And these four profiles are but a small sample of the wide range of quite distinct reflectance profiles that all have the same subjective effect on the human visual system. The fact is, our rather crude resources for processing chromatic information – namely, the three types of wavelength-sensitive cone cells, and the three types of 'color-opponency' cells to which they ultimately project – are simply inadequate to distinguish between these metamers. Any object boasting any one of them will look to be a qualitatively uniform yellow, at least under normal illumination.

These examples concern the color yellow, but a similar diversity of same-looking metamers attends every other color as well. If one had hopes for a smooth *reduction* of each of the commonsense colors to a uniquely corresponding reflectance profile, those hopes are here frustrated; first, by a real diversity of reflectance profiles corresponding to each visually distinguishable color; and second, by our apparent inability to characterize what *unifies* the relevant class of diverse reflectance profiles, appropriate to each visually distinguishable color, *independently* of appealing to the qualitative character of the visual sensations they happen to produce in the idiosyncratic visual system of the human brain. If that is the *only* way in which we can specify what unites the class of metamers specific to any color, then either we must resign ourselves to a deflationary *relational* reconstrual[2] of what common sense plainly takes to be *monadic* properties of material objects, or we must resign ourselves to the elimination of objective colors entirely, as Larry Hardin, coherently enough, recommends.[3]

---

[2] J. Cohen, "Color Properties and Color Ascriptions: A Relationalist Manifesto," *Philosophical Review* (forthcoming).

[3] L. Hardin, *Color for Philosophers: Unweaving the Rainbow,* exp. ed. (Hackett, 1993), 300 n. 2.

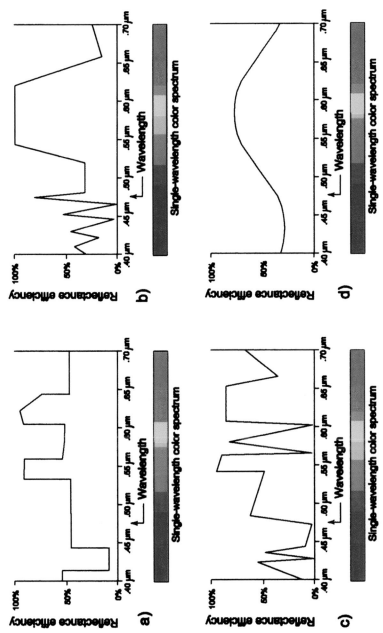

FIGURE 10.1. Four metamers for yellow

II. Reformulating the Problem

That the apparently unprincipled diversity of metamers poses a genuine problem for a reductive account of objective colors can be seen from a second perspective, one of central importance for understanding how the brain portrays the external world. A promising general approach to understanding how the brain – or any of its various subsystems – *represents* the external world posits the brain's development, through learning, of a variety of (often high-dimensional) *maps* of the objective similarity-structure of this, that, or the other objective feature-domain. Through extended experience with the relevant objective feature-domain, the relevant part of the brain can construct an internal map of that domain – of the range of possible faces, or the range of possible voices, or the range of possible reaching motions, or the range of possible colors, and so forth. Such internal maps represent the lasting or fixed structure of each external feature-domain, and they constitute the brain's *general* knowledge of the world's objective structure, that is, of the entire range of *possible* features that the world might display at any given time and place.

Once these conceptual resources are in place, the ongoing activity of the brain's various sensory systems will produce fleeting activations at specific *locations* within those acquired background maps, activations that code or index where, in the space of background possibilities comprehended by the map, the creature's current objective situation is located. For example, I am now looking at *my wife's* face; I am listening to *my wife's* voice; she is reaching for *a coffee mug*; and that coffee mug is *white*. In sum, I have a background conceptual framework – or rather, an interconnected system of such frameworks – and my sensory systems keep me updated on which of the great many possibilities comprehended by those frameworks are actualities here and now.

But the informational quality of such sensory indexings is profoundly dependent on the antecedent representational virtues of the background framework in which they fleetingly occur. The basic virtue of such background maps – as with any map – is a structural homomorphism between the map-as-a-whole, on the one hand, and the entire feature-domain that it attempts to portray, on the other. The family of *proximity relations* that configure the many map-elements of the brain's internal map[4] must have a relevant homomorphism with the family of *similarity relations* that

---

[4] Those landmark map-elements will be prototypical *activation patterns* across the relevant neuronal population.

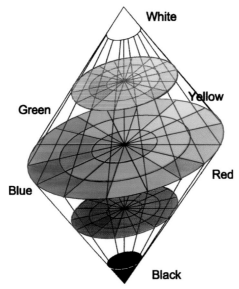

FIGURE 10.2. Our phenomenological color space

configure the many landmark features within the domain to be portrayed. Such homomorphisms or *second-order* resemblances, on this view, are the essence of the brain's representational achievements. One might call this account *Domain-Portrayal* Semantics, to contrast it with such familiar doctrines as *Indicator* Semantics or *Causal Covariation* Semantics.

I will not pause, in this essay, to detail the many virtues of this unified approach to how the brain represents the world's general or background categorical structure, and how it represents the world's local configuration here and now.[5] I sketch it here because it provides the background for a powerful contemporary objection to the reality of external colors in particular. "How," it may be asked, "does the peculiar and well-defined three-dimensional structure of the human phenomenological color space (see the spindle-shaped solid in Figure 10.2) *map onto* the objective space of possible electromagnetic reflectance profiles displayed by material objects? What is the internal structure of that objective target feature-domain in virtue of which the internal structure of our phenomenological color spindle constitutes an *accurate map* of that target domain?"

---

[5] For a broad account, look for my "Inner Spaces and Outer Spaces: The New Epistemology" (in progress).

The objector's questions here are, of course, semirhetorical. Their point is to emphasize the presumed fact that *there is no* objective structure that nicely configures the range of possible reflectance profiles displayed by material objects. Collectively, they form a noisy and unprincipled scatter of possibilities. At the very least, if there is some structure within that range of possibilities, it fails to answer in any way to the very specific and demanding structure displayed in our phenomenological color spindle. Objective colors, one might therefore conclude, are a Grand Illusion. The objective reality, concerning the surfaces of physical objects, is distinctly and importantly different from the naïve assumptions of common sense, and from the crude and misleading deliverances of our native sensory equipment.[6]

This is, at least potentially, a powerful argument *against* any common-sense view of colors as objective features of material objects. It appeals to the correct account of how objective feature-domains get represented in and by a brain, and it points to an apparently massive *failure* of the required second-order or structural representation in the specific case at issue. The unreality of objective colors is the presumptive consequence.

Nonetheless, I shall presume to resist this argument, because I think it rests on a false premise. Despite a negative first impression, there *is* a way to construe the initially opaque space of possible reflectance profiles so that its structural homomorphism with human phenomenological space becomes immediately apparent. Accordingly, our color space does map an objective reality after all, I shall argue, and thus the argument against color realism evaporates.

We understand one-half of this 'mapping conundrum' – namely, our phenomenological color space – quite well, both empirically and theoretically. The now-familiar Hurvich-Jameson opponent-process neural-network model of human color coding provides a compelling reconstruction of the empirical details of the spindle-shaped color solid of Figure 10.2. Figure 10.3*a* portrays the connectivity of that network, and Figure 10.3*b* portrays the wavelength sensitivity profiles of the three types of input cones. If one calculates the full range of possible activation-patterns across the three types of second-layer color-coding cells, given the details of the network's connectivity, that color-coding space turns out to have the shape portrayed in Figure 10.3*c*. Evidently, it has the same

---

[6] See again Hardin, *Color for Philosophers* (1993), 300 n. 2; see also E. Thompson, A. Palacios, and F. Varela, "Ways of Coloring: Comparative Color Vision as a Case Study for Cognitive Science," *Behavioral and Brain Sciences* 15 (1992): 16.

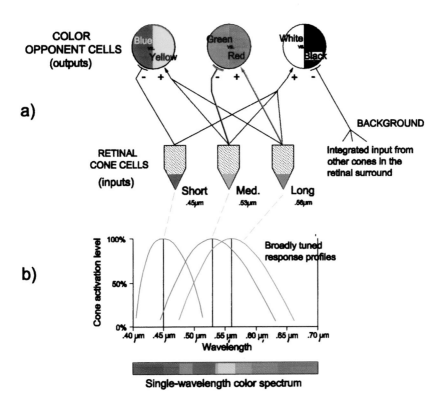

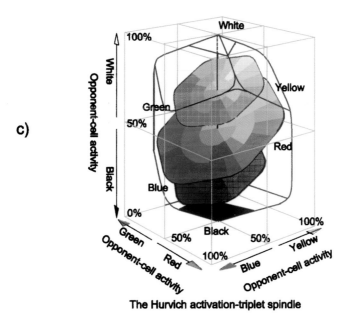

The Hurvich activation-triplet spindle

FIGURE 10.3. The presumed neuronal origins of our phenomenological color space

dimensionality, shape, and representational organization as the empirical color spindle, wherein lies its claim to explain the organization of our phenomenological color space.[7] This half of our problem – namely, the nature and ground of our internal map – is stable and more or less settled. It is the nature of the external reality being mapped that needs to be importantly reconceived.

III. Reconfiguring the Space of Possible Reflectance Profiles

The conventional way of representing any given reflectance profile that is located within the narrow window of the visible spectrum (see Figure 10.4a) positively *hides* an important feature of the range of possibilities therein comprehended. Perhaps the first hint of an alternative mode of representing those possibilities arises from the fact that the phenomenological color that corresponds to any narrowly monochromatic stimulus varies continuously across the visible spectrum, but it tends toward the *same* color – as it happens, a sort of deep purple/magenta – at each of the two opposite extremes: .40 μm at the extreme left, and .70 μm at the extreme right. It doesn't quite get there in either case, for no single wavelength of light will produce a sensation in the purple/magenta range. To get that (strictly nonspectral) range of colors, one needs simultaneous retinal stimulations at *two* places in the visible spectrum, toward its left and right extremes, respectively. But purple/magenta remains the missing color toward which each extreme tends. (Everyone since Newton has acquiesced in his constructing a continuous 'color wheel' in which the nonspectral purples are interposed to fill in the 'similarity gap' left open by the full range of single-wavelength stimuli.[8]) One's sense of rightful symmetry might therefore suggest that – as no more than an idle exercise, perhaps – one should pick up the planar figure in Figure 10.4a and roll it into a cylinder so that its right-most vertical edge makes a snug contact with its left-most vertical edge, as in Figure 10.4b. This converts the original planar space into a space that has no boundaries in the horizontal direction. It has boundaries only at the top and bottom of the space.

This trick turns the original reflectance profile itself, whatever its idiosyncratic ups and downs, into a wraparound configuration that admits

---

[7] For the details, see P. M. Churchland, "Chimerical Colors: Some Phenomenological Predictions from Cognitive Neuroscience," *Philosophical Psychology* 18, no. 5 (2005): 527–60 (Chapter 9 in this volume).

[8] See L. Hardin, *Color for Philosopher: Unweaving the Rainbow* (1988), 115, fig. III-1, for a portrayal of exactly where that nonspectral gap lies.

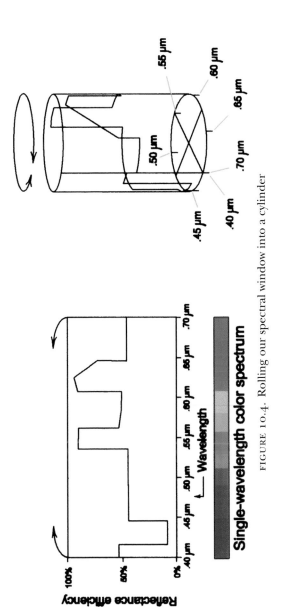

FIGURE 10.4. Rolling our spectral window into a cylinder

of an optimal approximation by a suitable planar cut through the now-cylindrical space. The locus of any such planar cut through the cylinder will always be an *ellipse* of some eccentricity or other (a circle in the limiting case of a planar cut that is orthogonal to the cylinder), as portrayed in Figure 10.5*b*.

The peculiar ellipse produced by a specific cut will be said to be an optimal – or, as I shall say henceforth, a *canonical* – approximation of the original or target reflectance profile when it meets the following two defining conditions:

1. The altitude of the ellipse must be such that the total area *A* above the canonical ellipse, but below the several upper reaches of the target reflectance profile, is *equal to* the total area *B* beneath the canonical ellipse, but above the several lower reaches of the target reflectance profile. (This condition guarantees that the *total area* under the target reflectance profile equals the *total area* under the approximating ellipse.)

2. The angle by which the ellipse is tilted away from the horizontal plane, and the rotational or compass-heading positions of its upper extreme, must be such as to *minimize* the magnitude of the two areas *A* and *B*. (This condition guarantees that the approximating ellipse *follows* the gross shape of the target reflectance profile, at least to the degree possible.)

A suitably situated, tilted, and rotated ellipse that meets these optimizing conditions, for a given reflectance profile, will be said to be the *canonical approximation* of that profile. Note that an indefinite variety of distinct reflectance profiles can share the very same ellipse as their canonical approximation. That clustering population, I shall propose, constitutes the class of metamers for whatever 'seen color' is produced by an object with a reflectance profile that displays their shared canonical approximation.

Equally important, for each and every individual reflectance profile, however jagged, there is a unique canonical approximation. (This is a consequence of sheer geometry, and of the definition provided earlier.) Note also that the canonical approximation for a given profile is an objective fact about that profile, and about the material object that possesses that profile. Its specification makes no reference to the human visual system, nor to the nature of its phenomenological responses to anything. The canonical approximation for the reflectance profile of a given material thing is an objective, mind-independent feature of that material thing. We

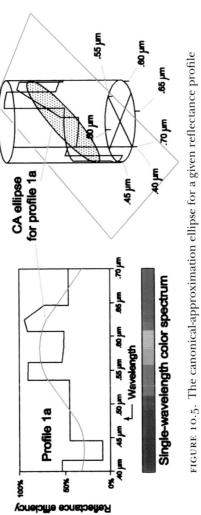

FIGURE 10.5. The canonical-approximation ellipse for a given reflectance profile

209

can safely be *realists* about whether a given reflectance profile has a specified ellipse as its canonical approximation (for short, its "CA ellipse"),
just as we can safely be realists about the reflectance profile thereby
approximated.

IV. How the Human Visual System Tracks CA Ellipses

Having identified such an objective, mind-independent feature of material objects, we might be tempted, straightaway, to identify any objective color with the canonical approximation of the relevant material
object's reflectance profile. But this is emphatically not my purpose. As
will emerge, my aim is the more narrowly focused aim of identifying
colors proper with the original, fine-grained reflectance profiles themselves, and not with their canonical approximations. But more of that in
a moment. For the present, I wish to point out that the changing activities
of the human visual system – as explored experimentally by generations
of psychologists since Munsell, and as portrayed in the familiar Hurvich-
Jameson network's[9] theoretical reconstruction of our phenomenological
color space (once again, see Figures 10.2 and 10.3*c*) – track the *canonical approximations* of the sundry reflectance profiles of various material
objects very effectively indeed. Let me illustrate, and let us begin by simply
examining the global structure of the entire space of *possible* CA ellipses.

The first thing to appreciate is that the space of possible CA ellipses
has three dimensions of variation: (1) the vertical position or altitude
of the given ellipse's center point within the reflectance-profile cylinder
of Figure 10.6*a*; (2) the degree to which that ellipse is tilted away from
being perfectly horizontal; and (3) the rotational position around the
cylinder of that ellipse's highest point. This three-space is clearly finite,
and it boasts the global shape portrayed in Figure 10.6*b*.

Note well its spindlelike or football-like configuration. The horizontal dimension (orthogonal distance away from the vertical central axis)
shrinks sharply to zero as the extreme top and bottom of the space is
approached. This reflects the fact that any CA ellipse in Figure 10.6*a*
will be progressively 'forced' into an increasingly horizontal position as
its altitude approaches the upper or lower extremes of the rolled-up
reflectance-profile space. Its 'tilt' must fall to zero as its altitude is forced
ever closer to the ceiling or the floor of that cylinder. Accordingly, the
horizontal dimension of the CA-ellipse space, which represents that tilt,

---

[9] L. M. Hurvich, *Color Vision* (Sunderland, MA: Sinauer, 1981).

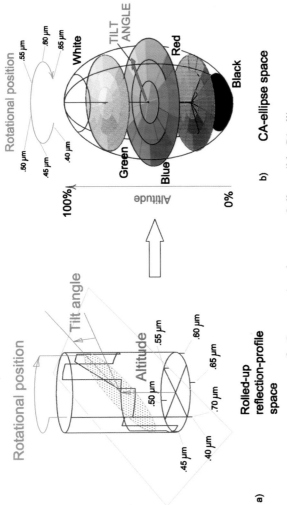

Rotational position

.50 µm .55 µm .60 µm
.45 µm .65 µm
.40 µm

TILT ANGLE

White

Green

Blue

Red

Black

100%

Altitude

0%

b)   CA-ellipse space

Tilt angle

Rotational position

Altitude

.55 µm
.60 µm
.65 µm
.50 µm
.70 µm
.45 µm
.40 µm

Rolled-up
reflection-profile
space

a)

FIGURE 10.6. Constructing the space of all possible CA ellipses

211

must also tend to zero at both the top and bottom of that space's vertical axis.

That tilt, recall, ultimately represents the degree to which an object's reflectance profile strongly favors some particular region of wavelengths over all of the other wavelengths in the spectral window .40 μm to .70 μm. And that dimension of variation corresponds very closely indeed to the dimension of color *saturation* displayed in the original *phenomenological* color solid of Figure 10.2. That dimension, of course, *also* shrinks to zero at the top and bottom extremes of that original space, wherein reside the hueless maximally bright white and the hueless maximally dark black, respectively.

To continue, the vertical dimension of the CA-ellipse space represents the altitude of a given ellipse's center point along the central axis of the reflectance-profile cylinder, which altitude ultimately represents the total area under the CA ellipse. That is, it represents the total energy of both the original reflectance profile itself and its CA ellipse (these, recall, are always the same). And that dimension of variation corresponds very closely indeed to the dimension of color *brightness* and *darkness* displayed in the original phenomenological color solid of Figure 10.2. That dimension bottoms out at maximal black and proceeds through its central axis to progressively lighter shades of gray, until it tops out at maximal white. In between those extremes, and away from the central axis toward the phenomenological space's outer surface, the various hues proceed from dark and weakly saturated versions of each (i.e., muddy versions), through maximally vivid or saturated versions at the equator, through progressively lighter and more weakly saturated versions of each (i.e., pastel versions) as we move up the color spindle. Here again, we confront another salient dimension of variation within our phenomenological space that corresponds very closely, this time, to the vertical dimension of the CA-ellipse space of Figure 10.6*b*. *Brightness*, evidently, is the objective feature therein represented.

Finally, there remains the dimension of angular position around the central axis of the CA-ellipse space. This dimension of variation reflects the angular position of the objective *high point* of the given CA ellipse in the rolled-up reflectance-profile space, which corresponds, in turn, to the *seen hue* within the phenomenological color solid of Figure 10.2. Evidently, a physical object's *hue* is the objective feature therein represented.

The CA-ellipse space (Figure 10.6*b*), let us remind ourselves, contains only *points*. (It is the rolled-up reflectance-profile space that contains the jagged profiles themselves and the wobbling ellipses that variously approximate them.) But that CA-ellipse space displays, immediately, exactly three dimensions, each of which corresponds to a salient dimension of our antecedently appreciated subjective phenomenological color space, which also has three dimensions. Moreover, each of these two spaces displays the same global *shape*: something close to a spindle or a football. Additionally, both spaces code the brightest objects at the upper tip of their spindles, and the darkest objects at the very bottom. Finally, both spaces code for the very same hues in their corresponding equatorial positions, in the same sequence as we proceed around that equator. Altogether, the internal structure of our subjective phenomenological color space provides an unexpectedly accurate *map* of the internal structure of the entirely objective CA-ellipse space.

Exactly *how* accurate is it? Topographically speaking, it is the answer to a color realist's prayer: three dimensions, exactly two of which present themselves in polar coordinates; the same global shape; and apparently all of the same betweenness relations. But how accurate is it *metrically*? It is very good, but not perfect. First, our phenomenological map is metrically deformed, somewhat, in the green/yellow/orange/red region, where the human L-cone sensitivity curve and the M-cone sensitivity curve substantially overlap each other.[10] This idiosyncratic feature of the human visual system for detecting color samenesses and differences makes us slightly hyperacute in that region. Because of this overlap, our color-processing system is here more sensitive to small changes in the dominant incident wavelength than it is to wavelength changes elsewhere in the optical window: in the short-wavelength or blue region, for example. The result is that the system counts smallish wavelength changes in the green-to-red region as equal in magnitude to somewhat larger increments of wavelength change elsewhere. You can see this metrical deformation directly by looking at the familiar rainbowlike color bars underneath Figure 10.1 *a* through 10.1 *d*. Those bars mark off equal increments of wavelength, but the 'seen colors' that correspond to them change only slowly in the blue region to the left, but rather more quickly in the green-to-red region toward the right.

Metrical deformations of some kind are a familiar feature of real-world maps. Think of the early-modern maritime maps made of the Americas. These were fairly accurate in the vertical direction, since the map-making ship's latitude was easily reckoned by the maximum nighttime altitude, above the horizon, of familiar stars. But they were notably inaccurate in

---

[10] See again Figure 10.3*b*.

their horizontal dimension, since the earliest expeditions had no accurate clocks, and thus no surefire way of determining their east–west or
longitude position as they made charts of their target coastlines. The west
coast of North America, for example, was occasionally misportrayed as
tilting almost 45 degrees to the left of its actual profile, all the way up to
Vancouver Island. Their inaccuracies aside, those maps were still maps. A
more exaggerated example of metrical deformation is that displayed in
any Mercator projection of Earth's surface, such as still grace the walls of
every grade-school classroom in America. As one approaches the north
and south extremes of such maps, their metrical (mis)representation of
east–west distances grows to absurd proportions. These gross metrical failings notwithstanding, the Mercator projection of Earth's surface remains
a paradigm example of a map, and a very useful one at that. Overall,
and metrically speaking, our color map is much more accurate than a
Mercator map of Earth.

Second, and as is to be expected, our internal phenomenological map
shows a nontrivial metrical deformation – this time in the vertical or
brightness dimension – in the areas toward the extreme left and the
extreme right of the optical window portrayed in Figure 10.1, for this is
where the absolute sensitivity of our S-cones and our L-cones falls to
zero.[11] As with measuring instruments generally, the accuracy of our
color-processing system plunges swiftly as one tracks its performance
at the extreme *limits* of its proprietary range of sensitivity. Specifically,
reflectance profiles with a substantial but isolated spike hard against
either end of the .40 μm to .70 μm window will get (mis)represented
as being essentially hueless, and as being much darker than they objectively are. In these narrow regions the visual system *fails* accurately to
track the objective tilt and altitude of a profile's CA ellipse, at least if the
relevant ellipse owes its proprietary configuration to a large reflectance
spike confined to that insensitive region. Such residual representational
failures are inevitable. They represent genuine, if minor, defects in the
human visual system for representing objective color, but they do not
represent any defect in the claim that the human visual system *does* represent objective CA ellipses. For it remains true that, these minor defects

---

[11] My thanks to an anonymous referee for forcing my attention toward this particular
imperfection in the human visual system's capacity to track similarities and differences
among CA ellipses. Its misrepresentations here are fairly minor and highly localized,
however, especially compared to those embodied in a Mercator projection, and they do
nothing to undermine the claim that our phenomenal space is a moderately faithful *map*
of CA-ellipse space *as a whole.*

aside, the phenomenological space in which our visual system codes its measurements plainly does constitute a recognizable map of the space of CA ellipses for objective reflectance profiles.

Moreover, and as if to make amends for its representational failures at, or very close to, the .40 μm/.70 μm boundary of the rolled-up reflectance-profile space of Figure 10.4*b*, the human visual system does indeed make effective discriminations of the actual configuration of CA ellipses whose high point lies anywhere close to that problematic boundary *if*, but *only if*, the reflectance profiles thereby approximated possess the bulk of their energies at *two* distinct wavelength spikes at some distance on *either side* of that discriminational 'dead point'. In fact, it is precisely such *two-headed* profiles that get coded, by the human visual system, with the familiar (but appropriately nonspectral) *purples!*

This idiosyncratic feature of human color coding has been familiar to color scientists for many years.[12] The CA-ellipse story of what it is that our visual system is coding for nicely accounts for this wrinkle. The fact is, it takes a reflectance profile containing two substantial energy peaks *straddling* that dead point (and little or no energy elsewhere in the spectral window) to yield a CA ellipse with a high point at that problematic boundary. The story also explains why maximally saturated purples are always so *dark*, relative to the saturated versions of all of the other colors. A maximally saturated purple requires a strongly tilted CA ellipse whose high point is located at the dead point boundary here under discussion. But that high point is doomed to be misrepresented by the H-J net, unlike high points elsewhere around the cylinder, for the more we concentrate the incident reflectance profile's two energy peaks toward the dead point, the feebler is the visual system's response. On the coding story here proposed, therefore, a maximally saturated purple is thus doomed to seem somewhat darker than any of the other saturated colors, at least to humans. And so it is.

All told, the structure of phenomenological space corresponds quite nicely to the structure of an antecedent space of specifiable objective features after all, namely, the space of possible CA ellipses. So long as we portrayed reflectance profiles as so many lines meandering across a flat and everywhere-bounded two-dimensional space, the manner in which they cluster into objective similarity classes was almost certain to remain opaque. But once we roll that space into a horizontally unbounded tube, such matters become much easier to see. My central proposal, therefore,

---

[12] See again fig. III-1 in Hardin, *Color for Philosophers* (1988), 115.

is that the objective physical feature that unites all of the reflectance-profile metamers[13] for any seen 'commonsense' color is the peculiar CA ellipse that they all share as their best approximation. And our phenomenological color space maps the range of *possible* CA ellipses very faithfully indeed, dimension for dimension, and internal location for internal location.

To see this directly, simply compare the space of possible CA ellipses portrayed in Figure 10.6*b* with the long-familiar space of possible color sensations portrayed in Figure 10.2 (and with the space of neuronal coding triplets portrayed in Figure 10.3*c*). Evidently, the differences are minor. First, the equator of the CA-ellipse space is not tilted up toward yellow, as is the equator of color-sensation space. This reflects, once again, the fact that the sensitivity curves of our three kinds of cone receptors are non-uniformly distributed across the human spectral window: the L- and M-cone curves overlap substantially. A saturated-yellow sensation (which requires a near-maximal external stimulation of *both* L- and M-cones) will therefore seem brighter than any other saturated color sensation. And second, the CA-ellipse space is plainly 'bulgy' or more egg-shaped than is the phenomenological spindle, as drawn in Figure 10.2. Figure 10.2 reflects the textbook orthodoxy of representing phenomenological color space as a double-*coned* spindle. But that portrayal is only a graphical convenience. Phenomenological color space, too, is more 'bulgy' than is conventionally portrayed in Figure 10.2, as has been known since Munsell first sought to portray it over a century ago. A more accurate portrayal would have it bulging outward somewhat, toward its top and bottom, which would bring its global structure even closer to the space of CA ellipses portrayed in Figure 10.5*a*. Finally, a mathematical reconstruction of the shape of the human color solid, based on the Hurvich-Jameson model network mentioned earlier (see again Figure 10.3*c*), also yields a space that is like the double-coned spindle of Figure 10.2, but is rather bulgier toward the top and bottom extremes.[14]

In all, our internal phenomenological color space is evidently a systematic homolog of the space of objective CA ellipses. It is a reliable map of the global structure of that external feature space. Moreover, our

---

[13] Well, *almost* all. Recall once more that the human visual system tracks CA ellipses increasingly *poorly* for reflectance profiles that display significant amounts of their energy in the narrow region of the 'dead spot', where .40 μm abuts .70 μm, as noted three paragraphs ago, and in footnote 11. This isolated failing can lead to (rare) profile-pairs that share the same objective CA ellipse, yet look slightly different to us.

[14] For the details of its derivation, see Churchland, "Chimerical Colors."

ephemeral sensory indexings within that background map (i.e., our fleeting color sensations themselves) are moderately accurate indications of *which* CA ellipse we might be confronting at any given moment. Finally, and most importantly, those CA ellipses evidently constitute the *resolution limit* with which the human visual system can access the objective and often jagged reflectance profiles of objects. That resolution limit is fairly coarse, to be sure, but there is something objective which is being reliably, if rather fuzzily, resolved: reflectance profiles across the entire spectral window. We call them colors.

(I should mention that the story just outlined is not the first attempt to find systematic similarities between the structure of our phenomenological color space and the structure of objective or physical color space. In a recent paper,[15] L. D. Griffin finds some notable similarities between the several 'symmetry axes' of the color spindle of Figure 10.2, and the 'symmetry axes' displayed in the less familiar CIE space for objective colors widely used in the lighting industry. I believe the parallels he finds are entirely genuine, if less comprehensive than the systematic structural isomorphism discovered on the present analysis. My only criticism is that he has chosen, as his representational target, the wrong space for objective color. The CIE space is a space for representing and analyzing *illuminants*, not reflectance profiles. It is a space for predicting the seen color that will result from mixing *light* at three utterly specific and canonical wavelengths, those corresponding to the focal λ-sensitivities of the human S-, M-, and L-cones. It is a perfectly good and useful space, but it does not address the reality of the objective colors of the vast majority of objects in our terrestrial environment, which are almost exclusively *reflectance* colors, not self-luminous colors. Moreover, it fails to represent the all-important dimension of objective lightness and darkness captured by the space of possible CA ellipses, as portrayed in Figure 10.6*b*. The CIE space has no room for black, for example, or for any of the darkish colors in the neighborhood of black. (The range of colors it comprehends corresponds most closely to a single horizontal cut through the equator of CA-ellipse space, a plane of constant brightness.) Nonetheless, Griffin's psychological/physical parallels are entirely welcome, for the colors of self-luminous bodies are as objectively real as are the more common reflectance colors. (More on self-luminous colors later, in Section VIII.)

---

[15] "Similarity of Psychological and Physical Color Space Shown by Symmetry Analysis," *Color: Research and Application* 26, no. 2 (2001): 151–7.

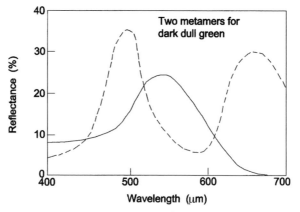

FIGURE 10.7. Hardin's metamers

V. Some Specific Tests

That the space of color sensation tracks (fairly closely) the space of CA ellipses is quite evident. But it is still a hypothesis – if a plausible one – that what unites the (uniform-illumination) metamers for any given humanly perceivable color is the CA ellipse that they severally share. (It is initially plausible because the coarse-grained resources of the human visual system typically *cannot tell the difference* between a given profile and its canonical approximation.) But let us quickly test the hypothesis against two salient examples of real metameric pairs, one drawn from Hardin[16] and the second drawn from Fraser.[17] The first example appears in Figure 10.7.

These two reflectance profiles are metameric pairs, according to Hardin, despite their evident differences. How do they compare with regard to their respective CA ellipses? To answer this question, I traced each of these profiles onto a separate transparency and rolled each into a cylinder. I then probed each profile (separately) with another rotatable cylinder slid inside it, a cylinder graduated with ellipses of varying tilt angles, until a 'closest match' was achieved, according to the criteria set out at the end of Section III. (The relevant areas were measured by integrating over a substantial number of narrow, vertically oriented rectangles.) This yielded a unique CA ellipse for each profile. The CA ellipse for the double-peaked profile has a peak at a rotational position

[16] Hardin, *Color for Philosophers* (1993), 47
[17] B. Fraser et al., *Color Management* (Berkeley, CA: Peachpit Press, 2003), 30.

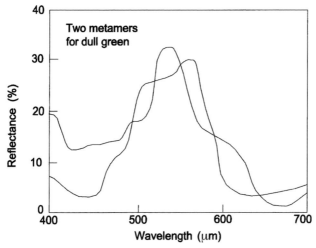

FIGURE 10.8. Fraser's metamers

$R = .52$ μm, a height $H = 14\%$, and a tilt angle $T = 17\%$ of maximum. The CA ellipse for the single-peaked profile has a peak at a rotational angle $R = .535$ μm, a height $H = 13\%$, and a tilt angle $T = 16\%$ of maximum.

The difference between these two CA ellipses is $\Delta R = 5\%$, $\Delta H = 1\%$, and $\Delta T = 1\%$. The difference is marginal, and both CA profiles (with peaks very close to .53 μm) will present as a dull and quite dark green – barely distinguishable, if they are distinguishable at all.

The next pair of metameric profiles also present to us as green, though a somewhat brighter and more saturated green than in the preceding example. The taller of these two profiles (Figure 10.8) was probed in the manner just described, and proves to have a CA ellipse of $R = .53$ μm, $H = 33\%$, and $T = 33\%$ of maximum. The second profile has a CA ellipse of $R = .53$ μm, $H = 29\%$, and $T = 35\%$ of maximum.

The difference between them is $\Delta R = 0\%$, $\Delta H = 4\%$, and $\Delta T = 2\%$. Once again, the differences are marginal – at or close to the limits of human discrimination.

Given the *systematic* match already noted between our phenomenological color space (Figure 10.2) and CA-ellipse space (Figure 10.6*b*), these singular matches should come as no surprise. But it is salutary to check out the hypothesis (that the class of same-seeming metamers for humans corresponds very closely to the class of reflectance profiles that share the same CA ellipse) against independent data.

VI. An Important Objection

There remains a possible objection to my claim that our phenomenological color space is a (fairly high-resolution) map of CA-ellipse space, and
thus is a (rather low-resolution) map of the range of objective reflectance
profiles. Hardin complains that our phenomenological color space displays an inescapable contrast between 'unmixed' colors (such as red or
blue) and 'mixed' colors (such as orange or purple), a contrast that is
completely absent in both the CA-ellipse space *and* the objective space
of possible reflectance profiles. How then can we identify colors with the
latter?

Let us agree, at least for the sake of argument, that both parts of
Hardin's claim are correct. This situation does nothing to undermine the
claim that our phenomenological color space accurately maps the space
of possible CA ellipses, for the structure of the latter *is* plainly reflected,
dimension for dimension, in the structure of the former. Hardin's antire-
alist argument here has got the 'onus of match' exactly backwards. It is not
incumbent on *the domain portrayed* to have every feature displayed by its
portraying map: maps can display all sorts of features that are incidental
to their role as effective maps (a common street map crumples easily and
weighs about an ounce, for example, in dramatic contrast to the urban
domain that it portrays). The contrast between mixed and unmixed phenomenal colors is just such an incidental feature – an artifact, presumably,
of the opponent-process architecture of our color system.[18]

What *is* required is that the relevant structure of the objective reality
(namely, the three dimensions of variation for a CA ellipse, as portrayed in
Figure 10.6*b*) finds itself reflected in some structural features of the map
that purports to portray that objective reality (namely, our phenomenological color spindle, as portrayed in Figure 10.2). In the present case,
that requirement is plainly met. That the map might have *other* features
that happen *not* to correspond to external structures is irrelevant.

My critique of Hardin's eliminativist position can perhaps be clarified
with the following parallel, drawn from another modality. The human
nervous system responds to temperature with two anatomically distinct
types of receptor neurons: one for registering temperatures *above* the
skin's temperature, and another for registering temperatures *below* it.
The first system produces a range of increasingly intense sensations all of
which are similar to one another, and so does the second. But the family

---

[18] See again the Hurvich-Jameson color-processing network of Figure 10.3*a*.

of sensations for warmth, on the one hand, and the family for coldness, on the other, are qualitatively quite distinct *from each other*. (No surprise, given that they arise from anatomically and physiologically distinct systems.)

Now, are we going to deny that objective temperature is identical with mean molecular kinetic energy on grounds that the objective scale of molecular kinetic energies embodies no such objective *qualitative* distinction between the regions above and below human skin temperature? Of course not. Nor should we hesitate, for similar (bad) reasons, to identify objective colors with reflectance profiles, on grounds that there is nothing in the domain of reflectance profiles that answers to phenomenological distinction between 'pure' and 'mixed' colors.

VII. What, After All, Are Colors?

Even so, it remains to discuss exactly how our familiar *objective colors* should be fit into this emerging framework. Simply identifying the familiar range of colors with the evident range of CA ellipses is a very poor option, since only a negligible proportion of material objects have a reflectance profile that is actually identical with the Platonic perfection of a CA ellipse. Any CA ellipse, of course, projects back onto the original reflectance-profile space as a perfectly smooth, one-cycle *sine wave* of some altitude, amplitude, and left/right location within that space (see again Figure 10.5). But most objects will have a much noisier reflectance profile than a perfect one-cycle sine wave: the meandering metamers still dominate the reflectance profiles we actually encounter in the real world. Accordingly, identifying the various colors with the various reflectance profiles displayed by perfect CA ellipses would have the consequence that almost nothing in the world is colored.

A much better option is to identify the full range of objective colors with the full range of objective reflectance profiles – both the relatively rare perfect CA ellipses *and* the multitude of metameric profiles that severally cluster around them – and then acknowledge that we humans are able, with our native visual equipment, to perceive and discriminate those highly various reflectance profiles *only at a rather low level of resolution*. As we noted earlier, the CA ellipse constitutes the *limit of resolution* at which humans can discriminate sameness and differences between objective reflectance profiles. In particular, we are typically unable to discriminate between any of the many metameric reflection profiles that share *the same* ellipse as their canonical approximation. These mutually clustered metameric profiles will typically present themselves, to the casual human eye, as *the same color*, despite the residual but real differences between them.

This situation, however, is entirely unremarkable. The human auditory system, to take a related example, is skilled at recognizing and discriminating power-spectrum profiles within the *acoustic* spectrum. We are good at recognizing and discriminating the distinct voices of people familiar to us, the distinct voices of the various types of musical instruments, birds, animals, and so forth. No one will deny that distinct types of sounds are identical with distinct power-spectrum profiles displayed in a propagating wave train, and no one will deny that our auditory skills reside in the cochlea's ability to respond to those various profiles in an appropriately discriminatory fashion.

And yet, our cochlea has a resolution limit as well. Clustered around the distinctive power-spectrum profile of a typical oboe's middle-A lies a multitude of possible "*acoustic* metamers," all of them different from one another in ways that lie beneath the capacity of my cochlea to resolve. Despite their differences, they will all sound the same to me. And so also for any other familiar sort of sound. At a certain point, and inevitably, our discriminatory powers simply run out. Such acoustic metamers for familiar *sounds* are as real, and as inevitable, as are the electromagnetic metamers for familiar *colors*.

But these undoubted facts about acoustic reality provide no grounds for irrealism or eliminativism about our commonsense ontology of sounds. Nor do the parallel facts, concerning electromagnetic metamers, provide grounds for irrealism or eliminativism about colors. Indeed, the ontological advantage, if any, should lie with colors. Sounds are ephemeral: a bird, a musical instrument, or an animal emits a sound only occasionally, and the sound fades (as $1/r^2$) to nothing as it promptly flees its point of origin. By contrast, a material body's electromagnetic reflectance profile is a quasi-permanent and stable property of that material body. It will change only if the molecular structure of the body's surface is modified in some way.

The stable solution then, to which we are thus attracted, is that the objective color of an object is identical with the electromagnetic reflectance profile of that object, within the window .40 μm to .70 μm. Our native ability to recognize and discriminate such profiles is limited to recognizing and discriminating the altitude, tilt, and rotation angle of the CA ellipse that approximates any given reflectance profile. But this native ability still gives us a highly reliable grip on an often-telling dimension of objective reality.

To be more specific, an object is a maximally saturated *red*, on this view, just in case its reflectance profile has a CA ellipse of altitude 50%, a maximum tilt, and a rotation position with the ellipse's highest elevation at .63 μm (see Figure 10.9*a*). An object is a somewhat dull *yellow*, on this view, just in case its reflectance profile has a CA ellipse of altitude 50%, a moderate tilt, and a rotation position with the ellipse's highest elevation at .58 μm (see Figure 10.9*b*). And so forth for every other objective color, no matter what its lightness, degree of saturation, and peculiar hue (see Figure 10.9*c–f*). We might think of this as the "wobbling penny" account of the space of possible reflectance profiles, for that is how the CA ellipse variously appears for diverse reflectance profiles.

This position has the consequence that two distinct objects, both of which are a maximally saturated yellow (or any other color), need not be exactly *the same* color, for they may sport distinct metamers included within the class *maximally saturated yellows*. They are both genuine instances of maximally saturated yellow, let us assume, but they may be *different* instances of what (as we now appreciate) is an interestingly diverse class. This description is, to be sure, a significant departure from our normal modes of speech, because common sense innocently *assumes* that there *are no* color differences underneath what our eyes can discriminate in broad daylight. But this naïve assumption must be let go. And the existence of diverse reflectance-profile metamers is precisely what demands its surrender. Even so, the existence of colors themselves, as objective features of objects, is not threatened. We simply have to acknowledge that there is slightly more to color than 'meets the human eye', even under the optimal conditions of broad daylight

Interestingly, that hidden diversity and sameness of objective colors is not *entirely* inaccessible to the human visual system. A simple trick will make such matters visually available, even to one who is color-blind. Given two shirt buttons, of apparently the same yellow color to normal vision, one can determine whether they are (1) *exactly the same* color (i.e., have identical reflectance profiles), or (2) merely have *distinct metameric variants* on a general yellow theme. We can do this by running both buttons, side by side, through the gauntlet of a rainbow projected on a wall (Figure 10.10*a*). If we project sunlight through a prism in an otherwise darkened room, as indicated, we will produce a fan of distinct monochromatic wavelengths to serve as diagnostic probes of each button's reflectance at any given wavelength of light. If the two buttons do share identical reflectance profiles, then their joint appearance to the human eye will *vary*, of course, as they are marched through the fan of distinct diagnostic

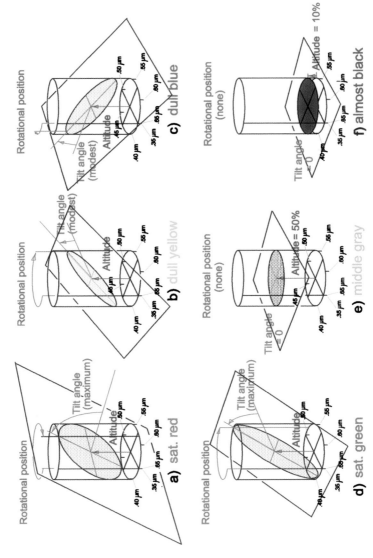

FIGURE 10.9. CA ellipses for six colors

224

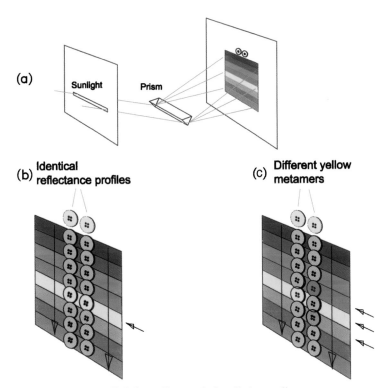

FIGURE 10.10. Rainbow diagnostic for distinct yellow metamers

illuminants. But at each position against the fan they will always display the same appearance *as each other* (Fig. 10.10*b*). By contrast, if the buttons have distinct metameric variants on yellow, then at one or more points in their journey across the rainbow they will appear *different from one another* (Figure 10.10*c*). They must. That they have distinct reflectance profiles entails that they will display differential reflectance behavior at some one or more points within the visible spectrum. Even a color-blind person will detect such discrepancies in their objective reflectance behaviors, since they will still present themselves, to him, as visible differences in apparent *gray-scale* brightness. In this way are the 'hidden' color metamers made visible, even to people who are color-blind.

Collateral or background information can also be a reliable guide to judging whether two same-seeming objects really have identical reflectance profiles, or merely share the same ellipse as their canonical approximation. If one is viewing two visually identical dark-red cherries, or two visually identical yellow bananas, for example, one can be

confident that the two cherries have genuinely identical reflectance profiles, and so also for the two bananas. For one can be independently confident that the two ripe cherries have identical molecular constitutions
at their surfaces, and so also for the two ripe bananas. Such identity in
molecular constitution physically guarantees identity in their reflectance
profiles. However, when background information suggests a quite different molecular constitution for two same-seeming objects – as with a purple
plum and a patch of purple paint – the distinct-metamers hypothesis will
have a better claim on the situation.

VIII. The Diversity of Objective Colors

It remains to highlight the contrast between the familiar *reflectance* colors, as characterized in the preceding pages, and the less-common *self-
luminous* colors, as displayed in a fire, a star, an incandescent bulb, or
an LED (light-emitting diode). The former is a matter of what profile
of light an object *reflects*; the latter is a matter of what profile of light
an object *emits*. Whatever common sense might think, these are entirely
distinct properties. One and the same object can simultaneously possess
incompatible 'colors' of each kind, as when a stove-top heating element
veridically presents its familiar reflectance color – a dark charcoal-gray –
when the kitchen lights are on; but when the lights are switched off and
the room is plunged into darkness, the element reveals its self-luminous
color of dull red (the relevant dial on the stove's control panel was set at
"low" all along). Though we could not see its self-luminous color in the
first condition – because the magnitude of the reflected light swamped
the comparatively faint emitted light – the darkened condition allows
that self-luminous color to become visible.

Self-luminous colors were an extremely rare occurrence in the evolutionary environment that gave birth to our current color vision. Only the
Sun, the stars, the occasional firefly, and the occasional forest fire *ever*
displayed a self-luminous color. Accordingly, and apart from the character of solar radiation as a background illuminant, the self-luminous
colors must have played a negligible role in the evolutionary selection of
the enabling mechanisms for human color vision. In modern society, of
course, the self-luminous colors have become commonplace. And in fact,
where metamers are concerned, the self-luminous colors are somewhat
better-behaved than are the reflective colors. The colors of a *thermally
incandescent* object almost always present a *smoothly varying* emittance
profile, whose peak magnitude is tightly tied to the object's absolute
temperature. And the self-luminous color of an object engaged in

spectral emission (i.e., in photon emission from electron-shell transitions) is almost always a matter of one or more *narrow spikes* of monochromatic light (as from an LED or a sodium streetlamp), which color is a reliable guide to the object's peculiar atomic constitution. Metamers are entirely possible here, as elsewhere, but in fact they are much less common for self-luminous colors than for reflectance colors.

The objective space for self-luminous colors (in the window .40 μm to .70 μm) is slightly but importantly different from the space for reflectance colors. In particular, the vertical dimension of the relevant cylinder represents not reflective efficiency (which tops out at 100%), but *emission intensity*, which has no upper limit. Nonetheless, our native representational space (specifically, the color-coding neuronal-activation space of Figure 10.3*c*) does its best to represent this relatively new range of possible *emittance* profiles, using its existing resources of brightness, saturation, and hue. But it here encounters an anomalous situation in that the *brightness* levels of typical self-luminous objects are much too high to be accounted for in terms of a 100% reflectance efficiency across the spectrum (i.e., as originating from a maximally reflective *white* object), and they often display a vivid *hue* (i.e., a nonwhite color) in any case. The human visual system responds by coding such anomalous (i.e., self-luminous) inputs at an appropriate place *on the ceiling* of our phenomenological color space, but *outside* the pointlike apex of the spindle-shaped volume that confines all of the representation points for the less dramatic reflectance colors (Figure 10.11).

These are 'impossible' positions, so far as the reflectance colors are concerned. (No *reflectance* color can be as bright as the brightest possible white and yet be something other than white.) But by that very fact, those unusual ceiling positions serve as reliable diagnostic positions, in our preexisting neuronal-activation space, to indicate the presence of a *self-luminous* object, and to indicate its peculiar hue and saturation. An information-processing system that was shaped by evolution to recognize and discriminate one kind of color turns out to be able to recognize and discriminate a second kind of color as well, and to do so in a manner that can reliably distinguish both.

Precisely because they are typically coded, by the visual system, *outside* the normal phenomenological color spindle, the self-luminous colors typically stand out like beacons against the darkness.[19] Their typical representational space is the *two-dimensional ceiling* of the opponent-cell

---

[19] Evidently, there is plenty of room, inside the human activation space for color-coding neurons, for coding vectors that lie *outside* the confines of the familiar color spindle (see

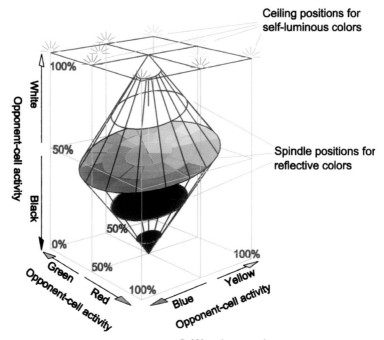

Ceiling positions for
self-luminous colors

Spindle positions for
reflective colors

FIGURE 10.11. Self-luminous colors

activation space. Save for a single point at its center, this space is entirely
distinct from the three-dimensional volume of the familiar spindle-
shaped solid for representing the reflectance colors. But it, too, maps
moderately well onto the objective range of its proprietary properties,
namely, the various *emittance profiles* displayed by self-luminous bodies.
(Exactly how the visual system discriminates between the different *brightness* levels of the self-luminous colors – note that the ceiling of the
opponent-cell space has only two dimensions – is a matter still unclear on
the present account. But I shall leave its pursuit for another occasion.)

I conclude that there are at least *two* quite distinct kinds of objectively
real colors, the reflectance colors and the self-luminous colors. The objective structure of each domain of properties becomes evident if one rolls
the window of the visible spectrum into an abstract cylinder, and then
examines the space of possible planar cuts through that cylinder, as providing the best approximation of the fine-grained details of the possibly
noisy power-spectrum profile currently portrayed around its surface. For

again Figure 10.2). These rogue coding vectors – representing "impossible" colors – are
explored at length in Churchland, "Chimerical Colors."

reflectance colors, the space of possible planar cuts (i.e., the space of possible CA ellipses) is homomorphic with the spindle-shaped solid of our phenomenological space. And for the more ostentatious self-luminous colors, the space of their possible hues and saturations is homomorphic with the space of possible activations within the otherwise unused two-dimensional *ceiling* of the opponent-cell activation space. The colors, of both kinds, are thus entirely real, and are as objective as you please. We can see them both, for in most cases we can recognize and discriminate both reflectance profiles and emission profiles, and discriminate the one from the other. The only disappointment here is both negligible and inevitable: some of the fine-grained structure of those profiles lies beneath the resolution of our native visual system. But this is no argument for irrealism about those profiles (of course), nor is it an argument for irrealism about colors. For that is precisely what the two kinds of colors *are*: reflectance profiles and emission profiles, respectively.

IX. Comparison with a Related View

The account of objective colors here defended shares many of the same motivations and some of the positive substance of the realist account of colors recently urged by Byrne and Hilbert.[20] But some important differences stand out, and I will close by bringing several of them to your attention. First, and perhaps least, those authors propose a new notion of a 'determinate color' (e.g., 'determinate red') whose extension is exactly the set of metameric reflectance profiles between which the human visual system is unable to distinguish. By contrast, I propose to identify distinct colors with distinct reflectance profiles and then embrace the consequence that the human visual system cannot distinguish all color differences since it cannot distinguish all reflectance-profile differences. We see color similarities and differences only down to the resolution limit defined by the range of possible CA ellipses. There are color similarities and differences beneath what we can detect just by looking. Our existing color vocabulary, therefore, comprehends only a coarse partitioning of the objective reality. Nevertheless, that partitioning is still objective in character and is highly useful as a guide to many of the causal properties of material objects.

---

[20] A. Byrne, and D. R. Hilbert, "Color Realism and Color Science," *Behavioral and Brain Sciences* 26 (2003): 3–21.

Second, and much more importantly, Byrne and Hilbert acquiesce in the received wisdom that the family of metamers for any commonsense color displays no unifying intrinsic feature specifiable in purely physical terms. They should not have acquiesced to this claim, because we can indeed specify, in terms "of interest to a physicist," the feature that unites the family of metamers for a given commonsense color: they all share the identical reflectance-space ellipse as their canonical approximation. Moreover, this shared objective feature is precisely what gets mapped within the human subjective or phenomenological color space. Accordingly, we can see *how* the space of human color sensations counts as a structurally accurate map of an objective domain of properties, a real achievement by our visual system that remains either denied or unrecognized on their view.

Finally, Byrne and Hilbert attempt to salvage a *unitary* conception of color by attempting to knit together the distinct features of reflectance, emittance, and transmittance into a single and deliberately more general notion of *productance*. By contrast, the view of this essay tends in exactly the opposite direction. I claimed earlier that reflectance colors are a family of properties genuinely distinct from the family of self-luminous colors. And I will say the same for a third family of 'color' properties: the various profiles of *transmittance* displayed by transparent and translucent objects such as colored glass, colored liquids, and some gemstones. All three types of color are features to which the human visual system gives us some nontrivial perceptual access, and in each case, that perceptual access involves our capacity to distinguish the power-spectrum profiles of the electromagnetic radiation arriving to our eyes. But the three types of objective properties themselves (reflectance, emittance, and transmittance) are radically distinct, from the point of view of informed physics. It would be folly to try to conflate them all into a single notion, especially when we already enjoy a perceptual system that allows us to spontaneously recognize the distinctions between them on a fairly reliable basis.

To add further force to this argument, the varieties of objective 'chromatic phenomena' do not end with the three types just mentioned. We must reckon also with the range of 'scatterance colors', as instanced in the 'blue' of the daytime sky. There are also 'interference colors', as displayed, for example, in various thin films such as an oil slick floating on water. There are also 'refractance colors', as get displayed when sunlight hits a prism, a many-faceted gemstone, or a spray of spherical water droplets. These three additional kinds of color *also* involve importantly *different* intrinsic properties, features, or mechanisms for interacting with light, each in its own characteristic way.

If we want to respect the impulse toward objectivity implicit in the commonsense conviction that the colors are real features of the external world, we should respect the lessons of modern physics that 'color properties' come in a substantial variety of objectively distinct families. We are merely stuck with a *single perceptual modality* – a trichromatic visual system – with which to ply access to all of them. But that is no grounds for conflating the distinct types of objective color themselves. They are importantly different from one another, even in the details of their visual appearance. Withal, and however various, those distinct types of color remain as real and as objective as you please, despite what commentators from Locke to Hardin have too hastily insisted. The colors – all six distinct families – deserve to be welcomed back into the fold of objectively real properties. We just need to understand them a little differently.

11

Into the Brain

Where Philosophy Should Go from Here

The physical brain, of both humans and animals, has begun to give up its secrets. Those secrets have been locked away in a bony vault, encrypted in a microscopic matrix of 100 billion neurons and 100 trillion synaptic connections, for the entire history of our philosophical musings, with no more influence on the content of those musings than the influence exerted by the equally hidden secrets of the kidney, or the secrets of the pancreas. The winding path of our philosophical theorizing has been steered by other factors entirely. Those factors have been many and various, even glorious, and they have been precious for existing at all. But they have not included even the feeblest conception of how the biological brain embodies information about the world, or of how it processes that information so as to steer its biological body through a complex physical and social environment. In these dimensions, we have been flying blind for at least three millennia.

But our blinders here have begun to be lifted, and our ignorance has begun to recede. A new generation of techniques and machines of observation has given us eyes to see into the encrypted details of neuronal activity. A new generation of scientists has given us a self-critical community of determined empirical researchers. And a new generation of theories has given us at least an opening grip on how the brain's massive but microscopic matrix might perform the breathtaking feats of real-time cognition that so compel our philosophical attention. My aim in this short paper is to outline the various ways in which the maturation of the cognitive neurosciences is likely to throw light on an unprecedented variety of issues of central and historical importance to philosophers in particular, issues near and dear to all of us, issues that have long defined our field. The overall impact of the maturing neurosciences, most will politely allow, is likely to be substantial. But most philosophers, I'll wager, expect the impact on philosophy to be relatively minor, if they have any expectations on the matter at all. How mistaken they are is the topic of this short paper.

Let me begin in what may be an unexpected place: semantic theory. How does the brain *represent* the enduring structure of the world in which it lives? The emerging answer, it seems, is surprisingly Platonic. The brain slowly develops, by a process to be discussed shortly, a high-dimensional *map* of the abstract categories, invariant profiles, and enduring symmetries that provide the unchanging *background structure* of the world of ephemeral processes. More accurately, the brain develops a substantial number of such maps, each one of which represents a specific domain of contrasting but interrelated universals, such as the domain of colors, the domain of voices, the domain of shapes, the domain of motions, the domain of animals, and so forth. Each map contains an appropriate family of *prototype* positions for each family of learned categories, and the assembled *proximity and distance relations* that configure those prototype positions within the map are collectively homomorphic with the assembled *similarity and difference relations* that configure the objective categories therein portrayed.

Unlike a street map, the brain's maps represent *abstract-feature domains* rather than concrete geographical domains (hence the allusion to Plato). But as with maps generally, representation is achieved not *atomistically* or one map-element at a time, but *holistically* or all map-elements together, by virtue of their collective internal structure, and by virtue of the homomorphism displayed between that internal structure on the one hand, and the similarity-structure of the relevant feature-domain on the other. The map is homomorphic with (at least a substructure of) the feature-domain being mapped. We might call this theory *Domain-Portrayal Semantics* to distinguish it from the various causal, covariational, indicator, teleological, and conceptual-role theories familiar to us from the contemporary philosophical tradition. Perhaps its closest cousin in that tradition is conceptual-role semantics (because both are holistic), but a contrasting feature of the present account is the fact that it has no dependence whatever on languagelike structures and structure-sensitive inferences. It embraces all cognitive creatures on the planet, language-using or no.

These internal maps of sundry external feature-domains (e.g., voices, shapes, motions) are embodied in the high-dimensional activation-spaces of the brain's many distinct neuronal populations, populations that typically number in the tens of millions of neurons. And just as any specific point on a two-dimensional highway map is specified by the simultaneous values of two variables – latitude and longitude – so is a specific point in an n-dimensional neuronal map specified by the simultaneous values of n variables – the current activation or excitation values of each of the n neurons in the representing population. As the number n climbs beyond the two dimensions familiar from a street map, the representational power of the n-dimensional map climbs proportionately. With the number n presenting in excess of tens of millions, for each of perhaps a thousand distinct maps within the brain, all of them interacting with each other, one starts to conceive a new respect for the representational powers of the biological brain, even for creatures well below us on the phylogenetic scale. As well, it now comes as no surprise that the bulk of one's background knowledge is deeply inarticulable.

If the story of the brain's grasp of the world's background structure is vaguely Platonic (plus or minus a prebirth visit to an abstract heaven), so also is the story for its unfolding grasp of the perceived here and now. Our perceptions of the ephemeral world are always and inevitably interpreted within the framework of whatever background maps we have already pieced together. Our perceptions make sense only against the background of our antecedently grasped concepts. For the primary function of our several sensory systems is continually to index *where*, in the space of abstract possibilities already comprehended, one's current empirical position resides. Our assembled sensory inputs, at any given moment, serve to activate a specific *pattern* of activation-levels across each of our waiting neuronal populations, a unique pattern for each map (remember: each map has its own abstract subject matter), a pattern that constitutes a "you-are-*here* pointer" to a specific possibility among the many background possibilities chronically portrayed in that map. We might call this the *Map-Indexing Theory of Perception*.

Very well, but a central problem for philosophy has always been, "How do we *acquire* our general knowledge of the world's categorical and causal structure?" Putting nativism aside – Plato's, Descartes', and Fodor's – we are left with a variety of empiricist stories that appeal to induction, hypothetico-deduction, falsification, Bayesian updating, or some combination thereof. But these are all 'category-dependent' forms of learning. They all require a determinate conceptual framework already in place, within which hypotheses can be framed, data can be expressed, and empirical reasoning can proceed. How such background frameworks are acquired in the first place is left unaddressed. Lockean/Humean stories concerning simple impressions and their residual copies – simple ideas – do attempt to fill this gap, but such stories are not empirically plausible, neither in their account of how 'complex' ideas are subsequently generated therefrom, nor in their account of how the alleged 'simple' ideas were generated in the first place.

If we ask, instead, how the *brain* develops its manifold maps of various abstract feature domains, developmental neuroscience already holds out the sketch of an answer. *Hebbian learning* is a mindless, subconceptual process that continually adjusts the strengths or 'weights' of the trillions of synaptic connections that intervene between one neuronal population and another, the very connections whose assembled weights *determine* the complex landscape of prototype regions that constitutes the abstract map embodied in the receiving population. Modify the synaptic weights and you modify the map.

More importantly, the Hebbian process of weight adjustment is systematically sensitive to *temporal coincidences* among the many axonal messages arriving, from an upstream population, to a given neuron in the receiving population. Specifically, if a cadre of connections, a subset among the great many connections to a given neuron, repeatedly bring their individual messages to the neuron *all at the same time*, then the weight of each connection in that united cadre is made progressively stronger. As neuroscience undergraduates are taught, "Neurons that fire together, slowly wire together." The receiving neuron thus gradually becomes a reliable indicator of whatever external feature it was that prompted the simultaneous activation of the relevant neurons in the sending population, the neurons whose axon tips embody the connections at issue. Moreover, since the salient features in any environment are those that display a repeated pattern of development over time (i.e., a distinct causal profile), the unfolding behavior of our Hebb-instructed receiving neuron can become an equally reliable indicator of a salient causal process out there in the world.

This sketch puts too much weight, perhaps, on the importance of a single neuron. Remember, there are thousands, even millions, of other neurons in the same population, who are presumably becoming sensitive, each in its own way, to some aspect or dimension of the same external feature-unfolding-in-time. It is the Hebb-trained population *as a whole* that eventually gains the important grasp of that target, and of the ways in which it contrasts with, or is similar to, a variety of other prototypical features-unfolding-in-time. In this way, presumably, does the mindless process of Hebbian weight adjustment gradually produce an internal map of an entire domain of abstract features, even if the infant creature's synaptic connections start off with random weight-values. The objective profile of our sensory inputs over time sculpts an internal representation of those statistics. That is, they sculpt a *map* of the world's chronic or enduring structure, both categorical and causal.

Thus does any creature acquire the skills of perception and causal recognition: it learns to activate appropriate points and paths through its background neuronal-activation spaces. Much the same process subserves its acquisition of bodily *motor* skills and the skills of manipulating its physical environment, as opposed to just passively observing it. Here, too, Hebbian learning sculpts representations: representations of the space of possible *actions*. *Practical* wisdom, it emerges, has the same sort of neuronal basis as does factual or theoretical wisdom, and in neither case do "laws" (in the latter case) or "maxims" (in the former case) play any fundamental role at all. Instead, one's level of wisdom is measured by the accuracy and the penetration of the high-dimensional maps one has constructed for the relevant abstract domains, both factual and practical. Plato, once again, would be pleased.

This holds for one's perceptual and navigational skills in the social and moral domains no less than in the various physical domains. Conventional wisdom has long modeled our internal cognitive processes, quite wrongly, as just an inner version of the public arguments and justifications that we learn, as children, to construct and evaluate in the social space of the dinner table and the marketplace. Those social activities are of vital importance to our collective commerce, both social and intellectual, but they are an evolutionary novelty, unreflected in the brain's basic modes of decision making. These have a different dynamics, and a different kinematics, entirely.

Upon reflection, this should come as no surprise. Baboon troops, wolf packs, and lion prides all show penetrating social perception and intricate social reasoning on the part of their members. And yet, lacking language entirely, all of their cognitive activity must be fundamentally nondiscursive. Why should humans, at bottom, be any different? Decision theorists, be advised. And moral philosophers. And jurists. And those whose job it is to study, and to try to repair, various cognitive and social pathologies. As with factual reasoning, practical reasoning and decision making is something we have but barely begun to understand. But the early lesson is that linguaformal models of practical cognition are catastrophically parochial.

To return to factual reasoning, the nature of cutting-edge scientific research looks interestingly different from the neuronal perspective as well. Making theoretical progress emerges as a matter of finding ever more penetrating and successful *interpretations* of the antecedently interpreted empirical data. It is not (usually) a matter of constructing fundamentally new maps for interpreting nature – that Hebbian process takes far too long. Rather, it is a process of trying to redeploy our existing conceptual resources in empirical domains *outside* the domain in which those concepts were originally acquired. Accordingly, Huygens reinterprets light as an instance of traveling waves. Newton reinterprets the orbiting Moon as a flung stone. Torricelli reinterprets the atmosphere as an ocean of air. Bernoulli reinterprets a gas as a swarm of ballistic particles. Each of these reinterpretations brought new insights and novel predictions in its wake. Theoretical science emerges as the critical exploration of revealing *models* and profitable *metaphors*, a process that involves the new use of old conceptual resources. Neural networks, as it happens, are entirely capable of modulating their normal conceptual response to any given class of stimuli. For the axonal projections that lead us stepwise *up* the brain's cognitive ladder(s) to ever more abstract maps embodied in ever more elevated neuronal populations, also project *downward*, in many cases, so as to allow cognitive activities at higher levels of processing to affect the ways in which familiar sensory inputs get processed at lower levels of interpretation. Brains, in short, can steer the way(s) in which they interpret the world, by making *multiple use* of the concepts that the very different and much slower process of Hebbian learning originally produced in them.

These downward-flowing or *recurrent* axonal projections are important for any number of reasons beyond the function just described. They are vital for producing prototypical *paths* (as opposed to mere points) in activation space, paths that represent causal processes-unfolding-in-time. And they are equally critical for mastering the recursive structures displayed in natural languages, for mastering the skills of arithmetic, the skills of geometry, the skills of logic, and the skills of music, all of which embody recursive or iterable procedures over well-formed structures. A brain with a purely feedforward architecture might do many things, but it could never master these skills. A brain with a recurrent architecture can.

Enough examples. We have gone through, or at least gestured toward, (1) a theory of concepts, with (2) an accompanying semantic theory; (3) a theory of perception, folded into (1) and (2); (4) a subconceptual theory of how any creature's conceptual resources are formed in the first place;

(5) a sublinguistic theory of motor knowledge and practical wisdom; (6) a sublinguistic account of social and moral knowledge; (7) a sublinguistic portrayal of practical reasoning and decision making; (8) a subdiscursive account of theoretical science; and (9) a non-Chomskyan account of our mastery of language and other recursive activities. Plainly, we are looking at a unified theoretical approach with an unusually broad reach.

There is much more to talk about, especially about the surrounding matrix of *human culture* and the manifold ways in which individual neural networks – that is, you and me – depend on and interact with that most blessed matrix. It is not a matrix of illusion (as in the silly movie by that name), but a matrix of acquired wisdom, an active framework that embodies many of the best achievements of the many earlier brains that also swam briefly in its nourishing informational embrace. This observation serves to illustrate that the neurocomputational perspective here paraded is not a narrow perspective, focused exclusively on the microarcana of individual brains. On the contrary, it is a multiscaled perspective that may finally allow us to construct a unified, and unblinkered, account of human cognition as it unfolds over the centuries. At the very least, it offers a systematically *novel* approach to problems that have always been central to our discipline. Concerning its future success ... I live in hope, as always. But now the reader will have some understanding of why.

I close with a historical parallel whose presumptive lesson will be plain to everyone. Recall our early attempts to understand the nature of Life, and the many dimensions of Health, prior to the many achievements of modern Biology, such as macroanatomy, cellular anatomy, metabolic and structural chemistry, physiology, immunology, protein synthesis, hematology, endocrinology, molecular genetics, oncology, and so forth. The medieval and premodern attempts, we can all agree, were downright pitiful, as were the medical practices that were based on them. But why should we expect our understanding of the nature of Cognition (cf. Life), and the many dimensions of Rationality (cf. Health), to be any *less* pitiful, prior to our making comparable achievements in penetrating the structure and the activities of the biological brain? Where should philosophy go from here? The answer could hardly be more obvious: into the brain.

Bibliography

Adolphs, R., Tranel, D., Bechara, A., Damasio, H., and Damasio, A. R. (1996). "Neuropsychological Approaches to Reasoning and Decision Making." In A. R. Damasio et al., eds., *The Neurobiology of Decision-Making*, pp. 157–80. Berlin: Springer-Verlag.

Anglin, J. M. (1977). *Word, Object, and Conceptual Development*. New York: Norton.

Bechara, A., Damasio, A., Damasio, H., and Anderson, S. W. (1994). "Insensitivity to Future Consequences Following Damage to Human Prefrontal Cortex." *Cognition* 50:7–15.

Bickle, J. (1998). *Psychoneural Reduction: The New Wave*. Cambridge, MA: MIT Press.

Byrne, A., and Hilbert, D. R. (2003). "Color Realism and Color Science." *Behavioral and Brain Sciences* 26:3–21.

Campbell, D. (1974). "Evolutionary Epistemology," in P. A. Schilpp, ed., *The Philosophy of Karl Popper*, pp. 413–63. La Salle, IL: Open Court.

Chalmers, D. (1996). *The Conscious Mind*. Oxford: Oxford University Press.

Cherniak, C., Changizi, M., and Kang, D. (1999). "Large-scale Optimization of Neuron Arbors." *Physical Review E* 59, no. 5: 6001–9.

Churchland, P. M. (1979). *Scientific Realism and the Plasticity of Mind*. Cambridge: Cambridge University Press.

Churchland, P. M. (1981). "Eliminative Materialism and the Propositional Attitudes." *Journal of Philosophy* 78, no. 2: 67–90.

Churchland, P. M. (1982). "Is *Thinker* a Natural Kind?" *Dialogue* 21, no. 2: 223–38.

Churchland, P. M. (1985). "Reduction, Qualia, and the Direct Introspection of Brain States." *Journal of Philosophy* 82, no. 1: 8–28.

Churchland, P. M. (1986). *Matter and Consciousness*. Revised edition, Cambridge, MA: MIT Press.

Churchland, P. M. (1988). "Perceptual Plasticity and Theoretical Neutrality: A Reply to Jerry Fodor." *Philosophy of Science* 55: 167–87.

Churchland, P. M. (1989a). *A Neurocomputational Perspective: The Nature of Mind and the Structure of Science*. Cambridge, MA: MIT Press.

Churchland, P. M. (1989b). "On the Nature of Theories: A Neurocomputational Perspective." In W. Savage, ed., *Scientific Theories*. Minnesota Studies in the Philosophy of Science, vol. 14, pp. 59–101. Minneapolis: University of Minnesota Press. Chapter 9 of P. M. Churchland (1989a).

Churchland, P. M. (1989c). "On the Nature of Explanation: A PDP Approach." Chapter 10 of P. M. Churchland (1989a). Reprinted in J. Misiek, ed., Boston Studies in the Philosophy of Science, Vol. 175, pp. 81–113. Dordrecht, Holland: Kluwer, 1995.

Churchland, P. M. (1989d). "Learning and Conceptual Change." Chapter 11 of P. M. Churchland (1989a).

Churchland, P. M. (1995b). *The Engine of Reason, the Seat of the Soul: A Philosophical Journey into the Brain*. Cambridge, MA: MIT Press.

Churchland, P. M. (1996a). "Fodor and Lepore: State-Space Semantics and Meaning Holism." In McCauley (1996), pp. 272–7.

Churchland, P. M., (1996b). "Second Reply to Fodor and Lepore." In McCauley (1996), pp. 278–83.

Churchland, P. M. (1996c). "The Rediscovery of Light." *Journal of Philosophy* 93, no. 5: 211–28.

Churchland, P. M. (1998). "Conceptual Similarity across Sensory and Neural Diversity: The Fodor/Lepore Challenge Answered." *Journal of Philosophy* 95, no. 1 (Jan.): 5–32.

Churchland, P. M. (1999a). "Densmore and Dennett on Virtual Machines and Consciousness." *Philosophy and Phenomenological Research* 59, no. 3 (Sept): 763–7.

Churchland, P. M. (1999b). "Review of *Reason, Regulation, and Realism*." *Philosophy and Phenomenological Research* 58, no. 4: 541–4.

Churchland, P. M. (2005). "Chimerical Colors: Some Phenomenological Predictions from Cognitive Neuroscience." *Philosophical Psychology* 18, no. 5: 527–60.

Churchland, P. M. (in preparation). "Inner Spaces and Outer Spaces: The New Epistemology."

Churchland, P. M., and Churchland, P. S. (1997). "Recent Work on Consciousness: Philosophical, Theoretical, and Empirical." *Seminars in Neurology* 17, no. 2: 179–86. Reprinted in P. M. Churchland and P. S. Churchland. *On the Contrary*, pp. 159–76. Cambridge, MA: MIT Press, 1998.

Churchland, P. S., and Sejnowski, T. J. (1992). *The Computational Brain*. Cambridge, MA: MIT Press.

Clark, A. (1997). *Being There: Putting Brain, Body, and World Together Again*. Cambridge, MA: The MIT Press.

Clark, A. (2000). "Word and Action: Reconciling Rules and Know-How in Moral Cognition." In R. Campbell and B. Hunter, eds., *Moral Epistemology Naturalized. Canadian Journal of Philosophy*. Suppl. vol. 26: 267–90.

Clark, Austen (1993). *Sensory Qualities*. Oxford: Oxford University Press.

Cohen, J. (forthcoming). "Color Properties and Color Ascriptions: A Relationalist Manifesto." *Philosophical Review*.

Cottrell, G. (1991). "Extracting Features from Faces Using Compression Networks: Face, Identity, Emotions and Gender Recognition Using Holons." In D. Touretzky, J. Elman, T. Sejnowski, and G. Hinton, eds., *Connectionist Models: Proceedings of the 1990 Summer School*, pp. 328–37. San Mateo, CA: Morgan Kaufmann.

Cottrell, G., and Laakso, A. (2000). "Qualia and Cluster Analysis: Assessing Representational Similarity between Neural Systems." *Philosophical Psychology* 13, no. 1: 77–95.

Cottrell, G., and Metcalfe, J. (1991). "EMPATH: Face, Emotion, and Gender Recognition Using Holons." In R. Lippman, et al., eds., *Advances in Neural Information Processing Systems*, vol. 3, pp. 1–7. San Mateo, CA: Morgan Kaufmann.

Cottrell, G. W., and Tsung, F. (1993). "Learning Simple Arithmetic Procedures." *Connection Science* 5, no. 1: 37–58.

Cummins, R. (1997). "The Lot of the Causal Theory of Mental Content." *Journal of Philosophy* 94, no. 10: 535–42.

Damasio, A. R. (1994). *Descartes' Error*. New York: Putnam.

Damasio, A. R. (1999). *The Feeling of What Happens*. New York: Harcourt.

Damasio, A. R., Tranel, D., and Damasio, H. (1991). "Somatic Markers and the Guidance of Behavior." In H. Levin et al., eds., *Frontal Lobe Function and Dysfunction*, pp. 217–29. New York: Oxford University Press.

Damasio, A. R., et al., eds. (1996). *The Neurobiology of Decision-Making*. Berlin: Springer-Verlag.

Davidson, D. (1970). "Mental Events." In L. Foster and J. Swanson, eds., *Experience and Theory*, pp. 79–101. Amherst: University of Massachusetts Press.

Dawkins, M. S. (1976). *The Selfish Gene*. Oxford: Oxford University Press.

Dawkins, M. S. (1982). *The Extended Phenotype*. San Francisco: Freeman.

Dennett, D. C. (1984). "Cognitive Wheels: The Frame Problem in Artificial Intelligence." In C. Hookway, ed., *Minds, Machines, and Evolution*, pp. 129–51. Cambridge: Cambridge University Press.

Dennett, D. C. (1991). *Consciousness Explained*. Boston: Little, Brown.

Dennett, D. C. (2005). "Two Steps Closer on Consciousness." In B. L. Keeley, ed., *Paul Churchland*, pp. 193–209. New York: Cambridge University Press.

Dennett, D. C., and Densmore, S. (1999). "The Virtues of Virtual Machines." In *Philosophy and Phenomenological Research* 59, no. 3: 747–67.

DYG (2000). "Evolution and Creationism in Public Education: An In-depth Reading of Public Opinion" (March). A national survey by DYG, Inc., 36A Padanaram Road, Danbury, CT 06811.

Edelman, S. (1998). "Representation Is Representation of similarities." *Behavioral and Brain Sciences* 21:449–98.

Elman, J. L. (1992). "Grammatical Structure and Distributed Representations." In S. Davis , ed., *Connectionism: Theory and Practice*. Vancouver Studies in Cognitive Science, vol. 3, pp. 138–94. Oxford: Oxford University Press.

Elman, J., Bates, E., et al. (1996). *Rethinking Innateness: A Connectionist Perspective on Development*. Cambridge, MA: MIT Press.

Feyerabend, P. K. (1962). "Explanation, Reduction, and Empiricism." In H. Feigl and G. Maxwell, eds., Minnesota Studies in the Philosophy of Science, vol. 3, pp. 28–97. Minneapolis: University of Minnesota Press.

Feyerabend, P. K. (1963a). "How to Be a Good Empiricist – A Plea for Tolerance in Matters Epistemological." In B. Baumrin, ed., *Philosophy of Science: The Delaware Seminar*, vol. 2, pp. 3–19. New York: Interscience Publications. Reprinted in B. Brody, ed., *Readings in the Philosophy of Science*, pp. 104–22. Englewood Cliffs, NJ: Prentice Hall, 1970.

Feyerabend, P. K. (1963b). "Materialism and the Mind-Body Problem." *Review of Metaphysics* 17: 49–66.

Flanagan, O. (1991). *Varieties of Moral Personality: Ethics and Psychological Realism.* Cambridge, MA: Harvard University Press

Flanagan, O. (1996). "The Moral Network." In McCauley (1996), pp. 192–215.

Fodor, J. A. (1974). "The Special Sciences," *Synthese* 28: 77–115.

Fodor, J. A. (1975). *The Language of Thought.* New York: Crowell.

Fodor, J. A. (1984). "Observation Reconsidered," *Philosophy of Science* 51: 23–43.

Fodor, J. A. (1988). "A Reply to Churchland's 'Perceptual Plasticity and Theoretical Neutrality.'" *Philosophy of Science* 55: 188–98.

Fodor, J. A. (1990). *A Theory of Content and Other Essays.* Cambridge, MA: MIT Press.

Fodor, J. A. (2000). *The Mind Doesn't Work That Way.* Cambridge, MA: MIT Press.

Fodor, J. A., and Lepore, E. (1992). "Paul Churchland and State-Space Semantics." Chapter 7 of *Holism: A Shopper's Guide.* Oxford: Blackwell. Reprinted in McCauley (1996), pp. 187–207. Oxford: Blackwell.

Fodor, J. A., and Lepore, E. (1996). "Reply to Churchland." In McCauley (1996), pp. 159–62.

Fodor, J. A., and Lepore, E. (1999). "All at Sea in Semantic Space: Churchland on Meaning Similarity." *Journal of Philosophy* 96, no. 8: 381–403.

Fodor, J. A., Garrett, M., et al. (1985). "Against Definitions." *Cognition* 8: 1–105. Amsterdam: Elsevier Science. Reprinted in E. Margolis and S. Laurence, *Concepts: Core Readings*, pp. 491–512. Cambridge, MA: MIT Press, 1999.

Fraser, B., et al. (2003). *Color Management.* Berkeley, CA: Peachpit Press.

Goldman, A. (1976). "Discrimination and Perceptual Knowledge." *Journal of Philosophy* 73:771–91.

Goldman, A. (1986). *Epistemology and Cognition.* Cambridge, MA: Harvard University Press.

Goodman, N. (1972). *Problems and Projects: Seven Strictures on Similarity.* Indianapolis, IN: Bobbs-Merrill.

Gorman, R. P., and Sejnowski, T. J. (1988a). "Analysis of Hidden Units in a Layered Network Trained to Classify Sonar Targets." *Neural Networks* 1:75–89.

Gorman, R. P., and Sejnowski, T. J. (1988b). "Learned Classification of Sonar Targets Using a Massively-Parallel Network." *IEEE Transactions: Acoustics, Speech, and Signal Processing* 36:1135–40.

Griffin, L. D. (2001). "Similarity of Psychological and Physical Color Space Shown by Symmetry Analysis" *Color: Research and Application* 26, no. 2: 151–7.

Hardin, C. L. (1988). *Color for Philosophers: Unweaving the Rainbow.*

Hardin, C. L. (1993). *Color for Philosophers: Unweaving the Rainbow.* Expanded edition. Hackett.

Hooker, C. A. (1995). *Reason, Regulation, and Realism: Toward a Regulatory Systems Theory of Reason and Evolutionary Epistemology.* Albany, NY: SUNY Press.

Hurlbert, A. (2001). "Trading Faces." *Nature Neuroscience* 4, no. 1: 3–5.

Hurvich, L. M. (1981). *Color Vision.* Sunderland, MA: Sinauer.

Huxley, T. H. (1866). *Elementary Lessons in Physiology.* Macmillan.

Jackson, F. (1982). "Epiphenomenal Qualia." *Philosophical Quarterly* 32, no. 127: 127–36.

Johnson, M. (1993). *Moral Imagination.* Chicago: University of Chicago Press.

Kitcher, P. (1982). *Abusing Science: The Case against Creationism.* Cambridge, MA: MIT Press.

Kripke, S. (1972). "Naming and Necessity." In D. Davidson and G. Harman, eds., *Semantics of Natural Language,* pp. 253–355. Dordrecht, Holland: D. Reidel.

Kuehni, R. (2003). *Color Space and Its Divisions: Color Order from Antiquity to the Present.* New York: Wiley.

Kuhn, T. S. (1962). *The Structure of Scientific Revolutions.* Chicago: University of Chicago Press.

Kuhn, T. S. (1974). "Second Thoughts on Paradigms." In F. Suppe, ed., *The Structure of Scientific Theories,* pp. 459–82. Urbana: University of Illinois Press.

Lakatos, I. (1970). "Falsification and the Methodology of Scientific Research Programs." In I. Lakatos and A. Musgrave, eds., *Criticism and the Growth of Knowledge,* pp. 91–196. Cambridge: Cambridge University Press.

Lehky, S., and Sejnowski, T. J. (1988). "Network Model of Shape-from-Shading: Neuronal Function Arises from Both Receptive and Projective Fields." *Nature* 333:452–4.

Lehky, S., and Sejnowski, T. J. (1990). "Neural Network Model of Visual Cortex for Determining Surface Curvature from Images of Shaded Surfaces." *Proceedings of the Royal Society of London* B240:251–8.

Leopold, D. A., O'Toole, A. J., et al. (2001). "Prototype-Referenced Shape Encoding Revealed by High-Level Aftereffects," *Nature Neuroscience* 4, no. 1: 89–94.

Levine, J. (1983). "Materialism and Qualia: The Explanatory Gap." *Pacific Philosophical Quarterly* 64:354–61.

Locke, J. (1689). *An Essay Concerning Human Understanding,* Book II, ch. viii.

Lockery, S. R., Fang, Y., and Sejnowski, T. J. (1991). "A Dynamical Neural Network Model of Sensorimotor Transformation in the Leech." *Neural Computation* 2:274–82.

MacIntyre, A. (1981). *After Virtue.* Notre Dame, IN: University of Notre Dame Press.

MacIntyre, A. (1999). *Dependent Rational Animals: Why Human Beings Need the Virtues.* La Salle, IL: The Open Court.

McCauley, R. N. (1996). *The Churchlands and Their Critics.* Oxford: Blackwell.

Nagel, T. (1974). What Is It Like to Be a Bat? *Philosophical Review* 83, no. 4: 435–50.

O'Brien, G. (1999). "Connectionism, Analogicity and Mental Content." *Acta Analytica* 22:111–31.

O'Brien, G., and Opie, J. (2006). "Notes Toward a Structuralist Theory of Mental Representation." In H. Clapin et al., eds., *Representations in Mind: New Approaches to Mental Representation,* in press. Westport, CT: Greenwood.

Pennock, R. T. (1999). *Tower of Babel: The Evidence against the New Creationism.* Cambridge, MA: MIT Press.

Popper, K. (1934). *Logik der Forschung.* Wien. Published in English as *The Logic of Scientific Discovery.* London: Hutchison, 1980.

Popper, K. (1972). "Conjectures and Refutations." In *Conjectures and Refutations,* pp. 33–65. London: Routledge.

Popper, K. (1979). *Objective Knowledge: An Evolutionary Approach.* Oxford: Oxford University Press.

Piaget, J. (1950). *Introduction a l'epistemologie genetique*, 3 vols. Paris: Presses Universitaires de France.

Piaget, J. (1965). *Insights and Illusions of Philosophy*. New York: Meridian Books.

Piaget, J. (1970). *Genetic Epistemology*. New York: Columbia University Press, 1970.

Quine, W. V. (1951). "Two Dogmas of Empiricism." *Philosophical Review* 60:**00–00**.

Quine, W. V. (1969). "Natural Kinds." *Ontological Relativity and Other Essays*, pp. 69–90. New York: Columbia University Press.

Rickless, Samuel C. (1997). "Locke on Primary and Secondary Qualities." *Pacific Philosophical Quarterly* 78:297–319.

Rojas, R. (1996). *Neural Networks: A Systematic Introduction*. New York: Springer-Verlag.

Rorty, R. (1965). "Mind-Body Identity, Privacy, and Categories." *Review of Metaphysics* 19:24–54.

Rosen, S., and Howell, P. (1987). "Auditory, Articulatory, and Learning Explanations of Categorical Perception in Speech." In S. Harnad, ed., *Categorical Perception: The Groundwork of Cognition*, pp. 113–60. Cambridge: Cambridge University Press.

Rosenberg, C. R., and Sejnowski, T. J. (1987). "Parallel Networks that Learn to Pronounce English Text." *Complex Systems* 1:145–68.

Roweis, S. T., and Saul, L. K. (2000). "Nonlinear Dimensionality Reduction by Locally Linear Embedding." *Science* 290, no. 5500 (Dec. 22): 2323–6.

Saver, J. L., and Damasio, A. R. (1991). "Preserved Access and Processing of Social Knowledge in a Patient with Acquired Sociopathy Due to Ventromedial Frontal Damage." *Neuropsychologia* 29:1241–9.

Schrodinger, E. (1944). *What is Life?* Cambridge: Cambridge University Press.

Sejnowski, T. J. (1988). "Computing Shape from Shading with a Neural Network Model." In E. Schwartz, ed., *Computational Neuroscience*, pp. 452–4. Cambridge, MA: MIT Press.

Sellars, W. (1963). *Science, Perception, and Reality*. London: Routledge.

Seung, H. S., and Lee, D. D. (2000). "Cognition: The Manifold Ways of Perception." *Science* 290, no. 5500 (Dec. 22): 2268–9.

Shepard, R. N. (1968). "Cognitive Psychology: A Review of the Book by Ulrich Neisser." *American Journal of Psychology* 81:285–9.

Shepard, R. N. (1980). "Multidimensional Scaling, Tree-Fitting, and Clustering." *Science* 210:390–7.

Strevens, M. (2003). *Bigger than Chaos*. Cambridge, MA: Harvard University Press.

Tenenbaum, J. B., de Silva, V., and Langford, J. C. (2000). "A Global Geometric Framework for Nonlinear Dimensionality Reduction." *Science* 290, no. 5500 (Dec. 22): 2319–23.

Thompson, E., Palacios, A., and Varela, F. (1992). "Ways of Coloring: Comparative Color Vision as a Case Study for Cognitive Science." *Behavioral and Brain Sciences* 15:16.

Tiffany, E. (1999). "Comments and Criticism: Semantics San Diego Style." *Journal of Philosophy* 96, no. 8: 416–29.

Toulmin, S. (1972). *Human Understanding*. Princeton, NJ: Princeton University Press.

Turing, A. (1950). "Computing Machinery and Intelligence." *Mind* 59:433–60.

Von Neumann, J. (2000). *The Computer and the Brain.* New edition. New Haven, CT: Yale University Press.

Zeki, S. (1980). "The Representation of Colours in the Cerebral Cortex." *Nature* 284:412–18.

Index